Video Engineering

Other McGraw-Hill Reference Books of Interest

Handbooks

BENSON • *Audio Engineering Handbook*

BENSON • *Television Engineering Handbook*

BENSON AND WHITAKER • *Television and Audio Handbook*

COOMBS • *Printed Circuits Handbook*

CROFT AND SUMMERS • *American Electricians' Handbook*

FINK AND BEATY • *Standard Handbook for Electrical Engineers*

FINK AND CHRISTIANSEN • *Electronics Engineers' Handbook*

HARPER • *Handbook of Electronic Packaging and Interconnection*

HICKS • *Standard Handbook of Engineering Calculations*

INGLIS • *Electronic Communications Handbook*

JOHNSON AND JASIK • *Antenna Engineering Handbook*

JURAN • *Quality Control Handbook*

KAUFMAN AND SEIDMAN • *Handbook for Electronics Engineering Technicians*

KAUFMAN AND SEIDMAN • *Handbook of Electronics Calculations*

KURTZ • *Handbook of Engineering Economics*

LENK • *Lenk's Audio Handbook*

LENK • *Lenk's Video Handbook*

MEE AND DANIEL • *Magnetic Recording Handbook*

SHERMAN • *CD-ROM Handbook*

TUMA • *Engineering Mathematics Handbook*

WILLIAMS AND TAYLOR • *Electronic Filter Design Handbook*

Consumer Electronics

BARTLETT • *Cable TV Technology and Operations*

BENSON AND FINK • *HDTV: Advanced TV for the 1990s*

BUNZEL AND MORRIS • *Multimedia Applications Development*

LUTHER • *Digital Video in the PC Environment*

MEE AND DANIEL • *Magnetic Recording, Volumes I-III*

PHILIPS INTERNATIONAL • *Compact Disc—Interactive*

WHITAKER • *RF Transmission Systems*

Dictionaries

Dictionary of Computers

Dictionary of Electrical and Electronic Engineering

Dictionary of Engineering

Dictionary of Scientific and Technical Terms

MARKUS • *Electronics Dictionary*

To order, or to receive additional information on these or any other McGraw-Hill titles, please call 1-800-822-8158 in the United States. In other countries, please contact your local McGraw-Hill office.

Video Engineering

Andrew F. Inglis
Consultant
Fair Oaks, California

McGraw-Hill, Inc.

New York St. Louis San Francisco Auckland Bogotá
Caracas Lisbon London Madrid Mexico Milan
Montreal New Delhi Paris San Juan São Paulo
Singapore Sydney Tokyo Toronto

To my wife Marie

Library of Congress Cataloging-in-Publication Data

Inglis, Andrew F.
  Video engineering / Andrew F. Inglis.
    p.   cm.
  Includes bibliographical references and index.
  ISBN 0-07-031716-X
  1. Television. I. Title.
  TK6630.I54 1993
  621.388—dc20   92-25055
                                                          CIP

    2 3 4 5 6 7 8 9 0   DOC/DOC   9 8 7 6 5 4 3

ISBN   0-07-031716-X

*The sponsoring editor for this book was Daniel A. Gonneau, the
editing supervisor was Paul R. Sobel, and the production supervisor
was Donald Schmidt. It was set in Century Schoolbook by
McGraw-Hill's Professional Book Group composition unit.*

Printed and bound by R. R. Donnelley & Sons Company.

Contents

ABOUT THE AUTHOR

Andrew F. Inglis is a consultant to the communications industry and recently retired as president and CEO of RCA American Communications. During his 30-year career at RCA, he also served as vice president and general manager of its Broadcast Equipment and Communications Systems Division. He is the editor of McGraw-Hill's *Electronic Communications Handbook* and the author of *Behind the Tube* and *Satellite Technology, an Introduction.*

Preface

Video Engineering describes the fundamentals of the principal video technologies that are employed in current television systems and that are planned for future systems. It can serve as an entry-level text for students who are preparing for a professional career in television research and engineering, as a reference source for scientists and engineers already engaged in these professions who need an introduction to the basic technologies outside their field of specialization, and also as a basic guide for engineers engaged in the design and operation of systems for terrestrial and satellite broadcasting, CATV, television program transmission, and television program production.

An author is faced with two problems in writing such a book. The first is the immense scope of the current video technologies, which would require many volumes for a detailed description. This problem has been resolved by emphasizing the fundamentals and providing references for the benefit of those desiring a more specialized and extensive description of specific subjects.

The second problem to be overcome is the rate at which new technologies are being introduced. The replacement of photoconductive vacuum tubes by solid-state imaging devices and the shift from analog to digital formats for high-definition TV (HDTV) signals are two examples. This creates the hazard of rapid obsolescence of a text that emphasizes the details of current equipment and systems. Again, the resolution is to emphasize the fundamentals that are necessary for an understanding of new developments, but which will be essentially unchanged by them.

This does not imply, however, that the descriptions of current equipment and systems should be neglected. Such descriptions are necessary, both to illustrate the application of these principles in a practical environment and to provide the reader with a knowledge of systems that are now in use and from which future systems will evolve.

The author is indebted to the many colleagues and competitors with whom he was associated in a 45-year career in the broadcasting and satellite industries. Their knowledge, critiques, and suggestions have been invaluable. He particularly wishes to recognize the contributions of Sidney Bendell, Walter Braun, John Christopher, Peter Dare,

Marvin Freeling, Charles Ginsburg, Anthony Lind, Arch Luther, Robert Neuhauser, Koichi Sadashigi, and Larry Thorpe. The excellent artwork was prepared by Thom Porterfield of Porterfield Design in Bend, Oregon.

Finally, he wishes to express his gratitude to his wife, Marie, whose support was invaluable during the long hours that she was separated from her husband by the word processor.

Andrew F. Inglis

Video Engineering

Television System Fundamentals

1.1 Introduction

This chapter summarizes the fundamental technologies of television systems as an introduction to later chapters, which describe their more detailed features.

Modern television systems are a marvel of ingenuity and complexity, but they are comparatively simple in concept, as shown in Fig. 1.1.

The camera lens focuses an image of the scene on the photosensitive surface of a sensor or *imager,* which converts an optical image into an electrical signal. Imagers were originally vacuum tubes—image orthicons, vidicons, and plumbicons—but more recently tubes are being supplanted by solid-state sensors such as charge-coupled devices (CCDs) (see Chap. 5).

The photosensitive surface of the imager is scanned with an electron beam (or by other electronic means), and the two-dimensional optical image is converted into an electrical signal in which variations in brightness are converted to their electrical analog, variations in voltage.

The camera signal passes to a processing amplifier that adds synchronizing pulses to time the picture tube scanning in synchronism with the camera scanning, and otherwise processes the signal for transmission.

The processed camera signal is amplified and modulates a radio-frequency (RF) carrier, using amplitude modulation (AM) for broadcasting or frequency modulation (FM) for transmission by microwave or satellite. (Transmission in digital form is also being used in an increasing number of applications; see Chap. 4.)

The transmission channel, which utilizes some combination of coaxial cable, fiber optic cable, microwave, satellites, and broadcast trans-

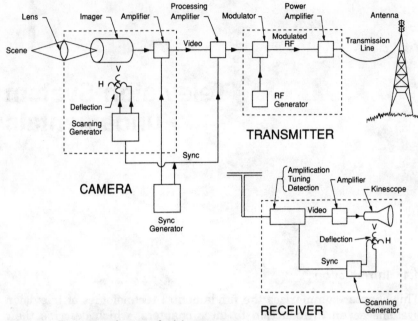

Figure 1.1   Basic monochrome television system.

mitters, is bandwidth-limited, either by the capacity of the RF spectrum or by technical factors. The bandwidth limitation of the transmission channel is a fundamental consideration in the design and performance of television systems.

In the receiver, the modulated RF carrier is amplified, the modulation is detected to recover the video signal, and the signal and synchronizing pulses drive the display device, usually a cathode-ray tube, which reproduces an image of the original scene.

Color systems are considerably more complex, but in principle they include only one more function, the addition of color to the final image.

1.2   Scanning

Scanning is used by all television systems to convert two-dimensional optical images into electrical signals. The invention of scanning was a major breakthrough and an indispensable first step in the development of television technology. Like many basic technical developments, it did not have a single inventor but appears to have occurred to a number of engineers simultaneously.

1.2.1 The scanning principle

The operation of the scanning principle in television is illustrated in Fig. 1.2. The optical image of the scene on the camera imager is scanned horizontally and vertically, and the imager converts the brightness variations along each scanning line to an electrical signal. Successive scanning lines are moved downward, one line at a time, by a vertical scanning signal to form a rectangular *raster*. The ratio of the width of the raster and the resulting image is known as the *aspect ratio*. Vertical scanning is repeated when the raster is complete.

At the receiver, the picture tube or other display device reverses the process. The variations in the electrical signal are reconverted to brightness variations, thus reconstituting the image. As a result of the retentivity of human vision, the eye perceives the rapidly repeated sequential display of scanning lines and rasters on the receiver tube as a continuous image.

1.2.2 Sampling

Scanning is a form of *sampling,* a process in which the value of a continuously varying signal is measured at regular time or distance intervals. The values of the samples are reconverted to a continuous function, either electronically or by the retentivity of the eye.

Two sampling procedures are employed in television scanning. Variations in vertical brightness are sampled by the scanning lines,

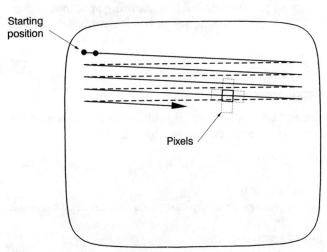

Figure 1.2 The scanning principle.

and the positions of objects in motion are sampled by scanning the entire raster repetitively.

The theory of sampling and its applications in television systems are described in Chaps. 2 and 4.

1.2.3 Limiting resolution

The *limiting resolution* of a television picture is one of its fundamental properties. It is expressed as the maximum number of alternate black and white lines that can be distinguished in a dimension equal to the height of the picture (both black and white lines are counted).

The vertical resolution, the number of horizontal black and white lines that can be distinguished or resolved in the picture height, is determined by the number of scanning lines. The horizontal resolution, the number of vertical lines that can be distinguished in a dimension equal to the picture height, is twice the number of cycles of the highest frequency in the video bandwidth in the time interval required for the scanning beam to traverse this dimension.

Limiting resolution and its significance in evaluating the performance of television systems is described more fully in Chap. 2.

1.2.4 Picture elements (pixels)

As a result of the scanning process, an image is portrayed as an array of rectangular picture elements or *pixels* (see Fig. 1.2). The vertical dimension of each pixel is the spacing between scanning lines. Its horizontal dimension is equal to the distance the scanning spot moves during one-half cycle of the highest frequency in the transmission-system bandwidth.

1.3 Bandwidth Requirements

The bandwidth required by a television signal is one-half the number of pixels transmitted per second. It is given by Eq. (1.1),

$$B_w = 0.8 F_R N_L R_H \tag{1.1}$$

where B_w = system bandwidth
F_R = number of *frames* or complete pictures transmitted each second
N_L = number of scanning lines
R_H = horizontal resolution

The numerical constant 0.8 is derived as follows:

$$B_w = (\text{cycles per frame})(F_R)$$

$$\text{Cycles per frame} = (N_L)(\text{cycles per line})$$

$$\text{Cycles per line} = \frac{(0.5)(\text{aspect ratio})(R_H)}{0.84} = 0.8R_H$$

The factor 0.5 is the ratio of the number of cycles to the number of lines resolved.

The aspect ratio in the equation is required because the horizontal resolution is measured in lines per picture height.

The factor 0.84 is the fraction of the horizontal scanning interval that is devoted to signal transmission. (The remainder is utilized for horizontal blanking.)

Choosing the optimum combination of B_w, F_R, N_L, and R_H was a major concern of the original system designers, since it involves a complex series of tradeoffs.

From the above it can be seen, that a wide bandwidth is required to resolve fine detail while maintaining a high enough picture repetition rate to avoid objectionable flicker. This explains the huge spectrum requirements of television systems.

1.4 Scanning Standards

The choice of the number of scanning lines requires a tradeoff between the conflicting requirements of low bandwidth, low flicker, and high resolution. For HDTV systems, *aliasing*, the formation of moiré patterns as the result of a beat between the sampling rate and signal frequency components (see Sec. 2.17.6), must also be considered.

The nature of the choice when moiré is not a problem is expressed by Eq. (1.1).

Spectrum space for broadcast systems is limited, and it is desired to keep the bandwidth as low as possible.

The frame rate F_R must be large enough to reduce flicker to an acceptable level. Also, it is desirable that it equal or nearly equal a multiple or submultiple of the primary power frequency so that "hum bars" resulting from imperfect filtering of the power source are stationary or slow-moving. This requirement resulted in the choice of approximately 30 frames per second in the United States and 25 frames per second in most European countries. Finally, in color systems, it is desirable that the frame rate be related to the frequency of the subcarrier used for transmitting the chroma information; see Chap. 3.

The number of scanning lines N_L and the horizontal resolution R_H should be as large as possible for maximum image sharpness and to minimize moiré effects. It is also desirable that the vertical and horizontal resolution be approximately equal.

1.4.1 Interlaced scanning

The difficulty in making the tradeoffs inherent in Eq. (1.1) is eased by the use of interlaced scanning (see Fig. 1.3). It permits the frame repetition rate, and hence the bandwidth, to be reduced by one-half with very little increase in flicker.

Each frame is divided into two *fields,* with the even and odd lines scanned on alternate fields. The perception of large-area flicker is based on the field rate, but the bandwidth requirements are based on the frame rate, which is one-half the field rate.

The use of interlaced scanning in television to reduce flicker is analogous to the use of a shutter in a film projector that exposes the film to two light flashes per frame.

To generate interlaced scanning, each frame must have an odd number of scanning lines, as shown in Fig. 1.3. This causes the first lines of alternate fields to begin in the center of the picture, so that the lines are interleaved. In the system employed in the United States there are 525 scanning lines in each frame and 262½ lines in each field.

As with all techniques employed in television to reduce the bandwidth

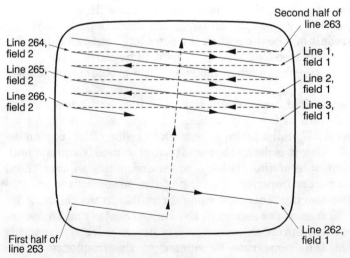

Figure 1.3 Interlaced scanning.

requirements, there is a price for the benefits of interlaced scanning. It is not always completely effective because of the tendency of the scanning lines to pair in the receiver, i.e., to fail to interlace. This is a receiver design problem that is described in Chap. 13. When pairing occurs, the effective number of scanning lines and the vertical resolution are reduced by one-half, thus eliminating the advantage of interlacing.

A more fundamental problem is small-area flicker. Fine vertical detail is repeated at the slower frame rate, and a sharp horizontal edge that is reproduced only by an odd or even field will appear to flicker.

Another problem with interlaced scanning is the greater visibility of moiré effects due to aliasing because the number of scanning lines is halved for moving objects.

Small-area flicker and moiré patterns are not particularly noticeable in most standard broadcast transmissions because the sharpness of the edges and the fineness of detail produced by the camera signal are not sufficient to generate visible spurious signals. They become a significant problem in HDTV systems, in which the images are sharper and the detail finer. As a result, the issue of interlacing has been raised again for HDTV.

1.4.2 Sequential or progressive scanning

The problems of interlaced scanning described in the preceding section can be greatly reduced or eliminated by the use of sequential or progressive scanning, in which every line is scanned on every field. The penalty, of course, is that twice the bandwidth is required. The probable result will be that sequential scanning will not be used for broadcast HDTV, but may be used in HDTV production systems or in wideband systems that employ alternative transmission modes, e.g., fiber optics.

1.5 Aspect Ratio

The aspect ratio, the ratio of the image width to its height, has a major esthetic effect on the appearance of the picture. The original motion picture aspect ratio was 4/3, that is, the picture width was ⁴/₃ the height. Since this ratio appeared to be attractive to the public, it was copied for television broadcasting.

With the passage of time, however, wide-screen motion pictures with aspect ratios in excess of 4/3 became increasingly popular for theatrical presentations. Most HDTV systems, again following the motion picture example, also employ wide-screen displays. There is no universally accepted standard aspect ratio for HDTV systems at this time, but a common choice is 16/9.

1.6 U.S. and Foreign Scanning Standards

The scanning standards that have been adopted by the United States and principal foreign countries are tabulated in Table 1.1. For reasons described in Chap. 3, these standards are slightly different for monochrome and for color.

1.7 Video Signal Waveforms

This section and the next give, a brief overview of the essential features of the waveforms that are currently used in television broadcasting and cable television. As described more fully in Chap. 3, the waveform for a color television signal consists of a luminance component that contains the image brightness information and a chrominance component or components that transmit the two color parameters, hue and saturation or color purity. The luminance component is almost identical to a monochrome signal, and it is this identity that makes color transmissions compatible, i.e., they can be received on either monochrome or color receivers. This chapter describes the luminance (monochrome) component signal. The chrominance component is described in Chap. 3.

TABLE 1.1 U.S. and Foreign Scanning Standards

	Color type	Fields per second	Frames per second	Lines per frame	Lines per second
United States					
Monochrome		60	30	525	15,750
Color	NTSC	59.94	29.97	525	15,734
England					
Monochrome		50	25	405	10,126
Color	PAL	50	25	625	15,625
Japan					
Color	NTSC	59.94	29.97	525	17,734
France					
Monochrome		50	25	819	20,475
Color	SECAM	50	25	625	15,625
Germany					
Color	PAL	50	25	625	15,625
Former USSR					
Color	SECAM	50	25	625	15,625

1.8   Luminance Signal Components

The waveform of the luminance components of a color signal is shown in Fig. 1.4. It has four sections: the picture waveform, blanking pulses, synchronizing pulses, and equalizing pulses.

1.8.1   Picture waveform

The picture waveform, which consists of the signal variations that correspond to variations in picture brightness, is transmitted during the active period of each scanning line. It has a black-positive polarity, i.e., the signal voltage is more positive in the black areas of the picture. This minimizes the visibility of noise, which is inherently greater in the black areas.

On a voltage scale, in which the tips of the synchronizing pulses are 100 percent, the picture signal varies from 7.5 percent for white areas to 70 percent for black. Its peak-to-peak amplitude is 62.5 percent of the amplitude of the composite video signal.

1.8.2   Blanking pulses

The blanking pulses cause the picture tube to go dark during the horizontal and vertical retrace times. They are slightly more positive than the picture signal from the blackest areas of the picture (the difference being called the setup), and they are sometimes described as "blacker than black." The horizontal blanking pulses occupy 14 to 18 percent of each line interval, while the vertical blanking pulses occupy 7.5 percent of the vertical scanning interval. In total, the blanking pulses occupy about 25 percent of the total frame time.

1.8.3   Synchronizing pulses

The synchronizing ("sync") pulses are superimposed on the blanking pulses, and their purpose is to maintain the scanning on the picture tube in synchronism with that in the camera. The leading edges of these pulses establish the timing of the picture tube scanning, and it is important that they be as sharp as possible to maintain precise timing in the presence of noise. If these edges are not sharp, noise can cause timing errors, although the high Q of the horizontal deflection yokes that are used in modern receivers makes the timing of horizontal scanning relatively immune to noise. The FCC standards require that the rise time for the leading edge of the pulses be no greater than 0.25 μs. This requires a transmission bandwidth of 2 MHz, or one-half the standard 4 MHz bandwidth of the video signal waveform.

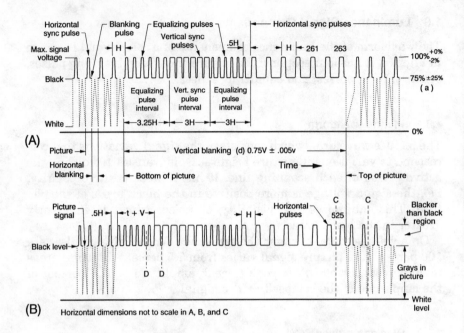

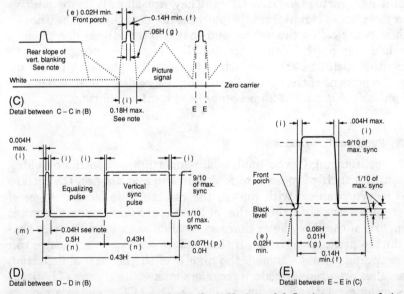

Figure 1.4 Luminance video signal waveform. Horizontal deflections not to scale in (A), (B), and (C).

1.8.4 Equalizing pulses

The purpose of the equalizing pulses is to prevent the pairing of interlaced lines on the receiver picture tube. The leading edge of the vertical sync pulses on odd-numbered fields coincides with the horizontal sync pulses, while the leading edge on even-numbered fields occurs in the middle of the lines. With the simple sync separating circuits that were used in early receivers, the residual effect of a horizontal pulse located one-half line before the beginning of a vertical pulse caused slight timing differences in the starting point of odd and even fields. This, in turn, caused pairing of the lines.

This effect was avoided by the use of two sets of six equalizing pulses with one-half the width and one-half the spacing of the synchronizing pulses. One set precedes the beginning of the vertical synchronizing pulse, and the other follows it. They equalize the conditions preceding the beginning of equal and odd-numbered vertical pulses and cause the trigger threshold to occur at the same location on the pulse. See Fig. 1.5.

1.8.5 DC component

Unlike audio signals, video signals have a dc component that depends on the average brightness of the picture. The signal generated by a very dark picture will have a larger dc component than that of a predominantly white picture. Since most electronic amplifiers do not transmit direct current, dc restorer circuits must be installed at critical points in the system, including the amplifier that drives the pic-

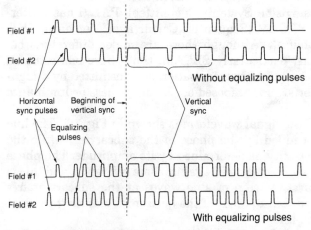

Figure 1.5 Correction of timing errors by equalizing pulses.

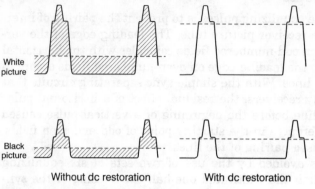

Without dc restoration   With dc restoration

Figure 1.6 Effect of dc restoration.

ture tube, to retain the proper brightness relationships from one scene to another. See Fig. 1.6.

Inexpensive monochrome receivers sometimes omit dc restoration to save cost. Without restoration, the average brightness, and hence the average picture tube current, remains the same for all scenes, which greatly reduces the demands on the power supply. This, however, causes serious distortion of the gray scale, which is unacceptable for color pictures. Accordingly, virtually all color receivers employ dc restoration.

1.9 Chrominance Signal Components

There are three systems of color television in wide use worldwide: NTSC (National Television Systems Committee), PAL (Phase Alternating Lines), and SECAM (Sequential Colour avec Memoire). They are described in Chap. 3. Although there are major differences between these systems, particularly SECAM, they share one common characteristic: the chrominance information is transmitted by a high-frequency subcarrier(s) superimposed on the brightness or luminance signal.

One line of an NTSC signal waveform is shown in Fig. 1.7. The hue of the color is determined by the phase of the subcarrier, while the color saturation, or purity, is determined by its amplitude. The phase of the subcarrier must be measured against some reference, and this is provided by the burst, a train of sine waves at the subcarrier frequency placed on the "back porch" of the horizontal blanking pulse at the beginning of the line.

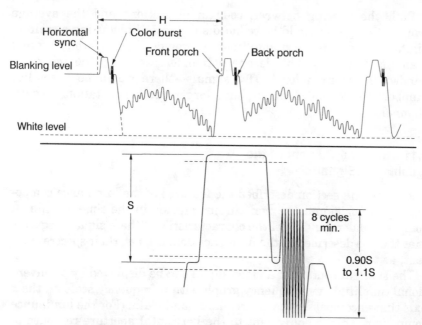

Figure 1.7 Color subcarrier.

1.10 Power Relationships in the Luminance Signal

The sync and blanking waveforms are very effective in fulfilling their intended functions, but they are extravagant in their use of power. On a voltage scale with the peak of sync at 1.4 and the blanking pulses at 1, the rms voltage of the blanking intervals over a complete frame is about 1.25. The rms voltage during the picture intervals depends on the picture content, but typically would be about 0.2. In this example, the average power level during blanking intervals would be proportional to 1.25^2 or 1.56, while the average picture power would be proportional to 0.2^2 or only 0.04. Although the blanking intervals occupy only 25 percent of the time, they utilize 97.5 percent of the transmitter power averaged over a complete frame. With a transmitting system having an effective radiated power of 100 kW, only 2.5 kW would be utilized for transmitting the picture signal.

Co-channel interference is also a problem with the high-powered sync and blanking pulses. They are the major source of interference because the lower-level picture signal from the desired station must override the higher-power sync and blanking pulses from adjacent interfering stations.

Both the spacing between co-channel stations and the average transmitted power could be reduced substantially by using a train of digital pulses of reduced amplitude to provide the synchronizing and blanking functions. This is one option for providing the increased bandwidth required for HDTV signals—there would be a smaller number of wideband channels, but a larger number of stations on each channel.

1.11   Spectral Content of the Luminance Signal

The preceding section described the features of the luminance components of color signals or monochrome signals in the time domain. A complete understanding of the characteristics of these signals requires that they be described in the frequency domain, i.e., their spectral content, as well.

The spectral content of a video signal can be displayed by a conventional amplitude vs. frequency graph with a frequency scale on the x axis that encompasses the entire video bandwidth. (For the luminance component, this is equivalent to the horizontal aperture response of the system that generates the signal, including the frequency response of the video channel; see Chap. 2.) This scale, however, is not sufficiently fine-grained to disclose an important property of video signals that can only be displayed with an expanded x axis.

1.11.1   Fine-grained spectral content

The fine-grained spectral frequency content of a television signal results from the fact that the signal contains three periodic components: one repeating at the line rate, one at the field rate, and one at the frame rate—nominally at 15,750, 60, and 30 Hz. Each component can be expressed by a Fourier series, a sequence of sinusoidal terms with frequencies that are multiples (harmonics) of the repetition rate—that is, 15,750, 60, and 30 Hz—and amplitudes that are determined by the waveform of the component. The summation of the amplitudes of the terms for the three periodic components over the entire video spectrum is the frequency content of the signal. The 30- and 60- and even the 15,750-Hz harmonic components are too closely spaced to be distinguished on a frequency scale of 4.2 MHz, and thus they are not disclosed on a conventional amplitude-frequency graph.

The nature of these harmonic components can be visualized by considering some special cases. (For the purpose of this analysis, the vertical blanking and synchronizing pulses are ignored.)

The simplest cases are all-white pictures and pictures consisting of vertical bars. Each scanning line is identical, and the frequency components are harmonics of the line frequency with amplitudes as determined by a Fourier analysis of the waveform, including the horizontal blanking and sync pulses. The number of harmonics is large, usually extending to the limit of the video bandpass. In a 4.2-MHz channel, it would be $(4.2 \times 10^6)/15{,}750$, or 266.

Another example is a picture in which the top half is white and the bottom half black. It generates two signal components that are sequential in time and periodic at the line rate, one for the white area of the picture and one for the black. The spectrum of the signal will consist of a series of frequency components harmonically related to the line frequency and with their amplitudes determined by a Fourier analysis of the white and black waveforms. In addition, however, the transitions from the black to the white signals will generate another series of frequency components that are harmonically related to the field frequency, the rate at which the black-to-white transitions are repeated.

The relationship between the line and field frequency components is analogous to that between the carrier and modulating frequencies in an amplitude-modulated signal. The field frequency components cluster around the line frequency harmonics like sidebands around a carrier, as shown in Fig. 1.8.[1,2] The field frequency components convey the information necessary to reproduce line-to-line changes in the brightness of the image, either in time as the result of motion in the scene or in space as the result of picture content. Their number and amplitudes depend on the rate of change of brightness in the vertical direction.

In a stationary scene with typical brightness variations, most of the vertical components are contained in a band ± 2 kHz on either side of the line rate harmonic, for a total bandwidth of 4 kHz. This is less than half the 15,750-kHz spacing between harmonics. If there are moving objects or unusually rapid vertical brightness variations in the image, the clusters broaden, but seldom so much as to occupy more than one-half the spacing between harmonics. An exception can occur when the vertical variations in brightness are extremely rapid. Under this condition, the clusters can broaden so far that the field rate components surrounding adjacent line rate harmonics overlap. This results in aliasing. In general, however, the standard television signal is prodigal in its use of spectrum space as well as power, and half of this space is unused.

In summary, the frequency spectrum of the luminance component of a color signal or of a monochrome signal consists of a series of harmonics of the line frequency, each surrounded by a cluster of frequency components separated by the field frequency. The amplitude of

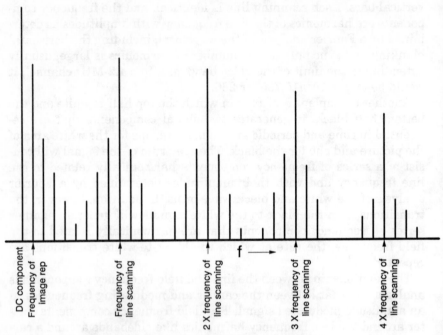

Figure 1.8   Fine-grained frequency spectrum of luminance signal.

the line frequency components is determined by the horizontal variations in brightness, while the amplitude of the field frequency components is determined by the vertical brightness variations.

1.11.2   Utilization of gaps in the video signal spectrum

The presence of gaps in the spectral content of luminance signals provides an important opportunity to transmit additional information
without requiring more scarce bandwidth. These gaps have been utilized for the following purposes:

1. *Compatible color systems.*   As will be described in Chap. 3, the frequency components of the color subcarrier and its sidebands, which
   also appear in clusters, are interleaved with the luminance frequency components. This minimizes the crosstalk between the luminance and chrominance signal components and makes it possible
   for the luminance channel of the receiver to utilize the full 4.2-MHz
   bandwidth of the video channel, even though it includes the color
   subcarrier. This is accomplished by the use of a comb filter in the

luminance channel that removes the chrominance channel components without disturbing the luminance.

2. *Signal-to-noise improvement.* A comb filter can also be used in the luminance channel of the camera to remove noise from the spectral regions between the line frequency clusters. Nearly half the noise can be removed by this technique without affecting the signal, leading to an improvement of about 3 dB in the signal-to-noise ratio.

3. *HDTV.* A number of techniques to utilize the spectrum space between clusters for analog HDTV systems have been proposed. See Chap. 14.

References

1. P. Mertz and F. Gray, "A Theory of Scanning and Its Relation to the Characteristics of the Transmitted Signal in Telephotography and Television," *Bell System Technical Journal*, Vol. 13, July 1934.
2. P. Mertz, "Television—The Scanning Process," *Proc. IRE*, Vol. 29, October 1941.

Elements of Television
Picture Quality

2.1 Overview

The technical performance of a television system is measured by the quality of the images it produces. An image is judged to be of high quality if it is a near-perfect replication of the original scene. Deliberate distortions of the image may be introduced for esthetic purposes—for example, defocusing the background areas of the scene—but the basic technical objective is to duplicate the appearance of the original as closely as possible.

The use of the word "appearance" in the preceding sentence indicates that the evaluation of television images is ultimately subjective. The television engineer, however, cannot be satisfied with exclusively subjective criteria of image quality. The design and operation of television systems require that these criteria be defined in objective and measurable terms. An important engineering task, then, is to develop objective picture quality yardsticks that coincide with the subjective perceptions.

The complexity of establishing objective quality criteria for television images is increased by the necessity of comparing television and photographic systems. Photography preceded television, and film quality has historically provided a basis for defining television image quality. A frequently stated goal in the design of broadcast television systems is to duplicate the quality of 16-mm motion picture film. Similarly, the goal of HDTV (high-definition television) systems is often described as the duplication of the quality of widescreen theatrical film.

Comparison of the quality of television and film images is difficult and cannot be rigorous. There are inherent differences between the television and photographic processes that make it almost impossible to duplicate precisely the appearance of a film image with a television system or vice versa. The television engineer nevertheless must establish practical image quality standards that are equivalent to those used for film, even though the comparison cannot be exact.

The challenge of establishing objective image quality criteria for television and of comparing the quality of inherently dissimilar film and television systems has stimulated an enormous amount of research by the photographic and television industries over the past decades. Practical definitions of the most significant measures of television image quality, together with their specifications and measurement procedures, have emerged from this research. A comprehensive listing of the criteria for evaluating the quality of television images, together with performance standards for both color and monochrome, is contained in a joint publication of the Electronic Industries Association (EIA) and the Telecommunications Association (TIA): "EIA/TIA Standard—Electrical Performance for Television Transmission Systems, EIA/TIA-250-C." Although this document is intended to set standards for transmission systems alone, it provides a useful basis for evaluating complete system performance. Since performance degradations are usually cumulative, the standards for the transmission system are more rigorous than those for the complete system.

This chapter begins with a brief summary of the most important quality criteria for monochrome television and the luminance component of color television. More detailed descriptions of these criteria are contained in later sections. The criteria for the chrominance components of color television are described in Chap. 3.

2.2 Basic Image Quality Criteria

Image quality criteria can be divided into two categories: basic criteria that apply to all imaging systems, and image defects, some of which are unique to television images. Basic criteria include image definition, gray scale, signal-to-noise ratio, and colorimetry.

2.2.1 Image definition

Definition is the degree of the "in-focus" appearance of the image. It is primarily measured by the sharpness of the transitions between dark and light areas of the picture. Although it is related to the resolution of fine detail in the image and is sometimes confused with it, image definition describes a different aspect of picture quality.

2.2.2 Gray scale

The gray scale specifies the image brightness as a function of scene brightness. It is sometimes called the transfer characteristic. It can be specified by three parameters: highlight brightness, the brightness of the brightest areas of the image; contrast ratio, the ratio of the brightness of the brightest areas of the image to that of the darkest; and gamma, the slope of the transfer characteristic of image brightness as a function of scene brightness, both expressed logarithmically.

2.2.3 Signal-to-noise ratio

In the broadest sense, "noise" is any unwanted signal, but in this usage it usually means random or thermal noise of a type that produces "snow" in a television image.

2.2.4 Summary

Definition, gray scale, and signal-to-noise ratio are fundamental criteria that apply to all imaging systems. They are also interrelated. For example, the perceived definition of an image is strongly influenced by its gray scale. For analytic purposes, however, it is usual to consider them separately. These basic quality criteria are described in detail in Secs. 2.5 to 2.16.

2.3 Image Defects

In addition to the basic criteria described above, image *defects* must be considered in evaluating television picture quality.

Image defects can be divided into three categories: (1) defects that are inherent in the system, such as flicker and aliasing, (2) defects such as lag and smear that result from deficiencies in the production of the video signal, and (3) spurious signals that are introduced by recording or transmission systems, such as hum and other forms of electrical interference, co-channel and adjacent channel interference, receiver-generated interference, and "ghosts." Sections 2.3.1 to 2.3.8 summarize the nature of these defects. See the Contents for the location of more detailed descriptions.

2.3.1 Flicker

Flicker (see Sec. 2.17.5) results when the frame repetition rate is not high enough to cause the eye to perceive a continuous image. In an interlaced system, the visibility of *large-area* flicker is determined by the field rate and that of *small-area* flicker by the frame rate.

2.3.2 Aliasing

Aliasing (see Sec. 2.17.6) is the production of spurious signals as the result of sampling in space or time.

2.3.3 Lag

Lag (see Sec. 5.4.7) is a measure of the rate of change of the video signal at a fixed point on the raster when the scene changes. Ideally, the signal voltage at any point would change instantaneously with a change in scene; if it does not, the result is a loss of resolution of moving objects, or in extreme cases, a trailing tail.

2.3.4 Geometric distortion

Distortion of image geometry, i.e., of the shape or size of objects, results from defects in the camera's optical system or nonlinearities in the scanning process.

2.3.5 Hum

The term "hum" is borrowed from audio engineering, and it describes interference caused by spurious power source voltages that have been introduced in the system. In its most common form, it results in a narrow horizontal "hum bar" that moves vertically through the picture at a rate equal to the difference between the field rate and the power source frequency.

2.3.6 Co-channel interference

Co-channel interference occurs when two television stations operating on the same frequency are received at the same location. The most common effect is "venetian blinds," alternate black and white horizontal bars across the picture that result from the "beat" between the two carriers.

2.3.7 Receiver-generated interference

A number of spurious signals can be generated by receiving systems. Two of these, local oscillator radiation and i.f. images, result from the use of the superheterodyne principle in receivers. A third, intermodulation interference, may result when the strong signals are received from two stations having carriers separated by the intermediate frequency of the receiver. These problems are particularly troublesome for UHF stations, but they are mitigated by the FCC's policy of channel assignments, which reduces the possibility of receiver-generated interference from the use of the superheterodyne principle. Another potential source of interference is intrachannel beat frequencies between the sound and picture carriers. The problems of receiver-generated interference are discussed in Chap. 13.

2.3.8 "Ghosts"

A "ghost" is a duplicate of the main image, slightly displaced from it, usually to the right, and much fainter. It is caused by multipath transmission, i.e., the signal radiated from the transmitter arrives at the receiver over two or more paths, usually a direct path and a reflection from a building or other large object. In spite of the enormous velocity of radio waves, the path-length differences between direct and reflected waves that are often encountered in practice are sufficient to produce noticeable displacements of the ghost image. Radio waves travel 63,000 ft during the time interval of one scanning line, and a 1000-ft path difference would produce a displacement of 1/63 of the scanning line length, or slightly more than 1/3 in on a picture tube with 20-in width.

2.4 Characteristics of Human Vision

The human eye is the final link in the perception of television images, and a consideration of its capabilities is critical in the design of television systems and the establishment of objective performance criteria. To state the obvious, there is no advantage in designing a system with image quality that exceeds the ability of the eye to perceive it.

The characteristics of human vision are exceedingly complex, and a detailed discussion of this subject is beyond the scope of this volume. The most significant characteristics for television system design are the perceptions of brightness and contrast, visibility of sharpness and detail, and flicker. These properties are described in this chapter. The perception of color is described in Chap. 3.

2.4.1 Brightness and contrast perception

Brightness is a subjective term that indicates the magnitude of the visual sensation produced by a source of light. *Contrast* is the ratio of brightnesses, usually expressed in decibels, between the darkest and lightest sections of the image.

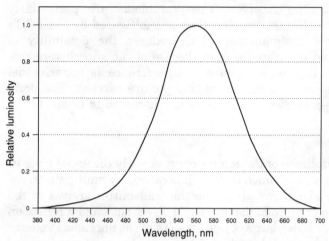

Figure 2.1 Luminosity function. (*From D. G. Fink,* Television Engineering, *McGraw-Hill, New York, 1952.*)

Luminance is the objective equivalent of the subjective sensation of brightness and is a measure of the *luminous intensity* of the surface of an extended light source. The luminous intensity is equal to the *luminous flux* or *lumens* emitted per unit solid angle. It is calculated by multiplying the energy or radiated flux expressed as a function of wavelength by a weighting factor, the *luminosity function*, that expresses the sensitivity of the eye to radiation of different wavelengths. Luminance is specified in *lamberts*, or *foot-lamberts*.

Figure 2.1* shows the luminosity function as standardized by the CIE (Comité International de l'Eclairage) in 1924.

Care should be exercised to avoid confusing *luminance* and *illuminance*. The former is a measure of the brightness of an area of an image. The latter is a measure of the illumination of a scene, for example, a television studio set, by an external light source. Illuminance is specified in *foot-candles*.

The visibility of an object in an image depends on the brightness contrast between different areas. As is the case with most physical sensations, the magnitude of this perception tends to be proportional

---

*Figure 2.1 is the *photopic* luminosity function, i.e., the response of the eye with normal levels of brightness. The *scotopic* function shows its response with very low levels of brightness. The peak of the scotopic function occurs at a shorter wavelength, approximately 510 nm.

to the brightness ratio rather than to the absolute brightness difference. This is expressed in *Weber's law:*

The increase in stimulus necessary to produce an increase in sensation of any of our senses is not an absolute quantity but depends on the proportion that the increase bears to the immediately preceding stimulus.

The perception of contrast between two areas also depends on the sharpness of the boundary. If the boundary is sharp, a much smaller brightness difference can be perceived.

2.4.2 Detail and image sharpness perception

One of the most significant characteristics of the eye for establishing the design parameters of a television system is its ability to discern detail and the sharpness of edges in an image. This attribute determines the eye's perception of two of the basic quality criteria for television images, definition and signal-to-noise ratio. It also has a direct effect on the bandwidth requirements for the system.

2.5 Image Definition

2.5.1 Image definition defined

The dictionary meaning of image definition, the distinctness of the outlines in the image, is also satisfactory as a technical definition. A less precise definition is the degree to which the image appears to be "in focus." Stated in still other terms, the definition of an image is a measure of the sharpness of the transitions or edges between its dark and light areas. In a high-definition system, these edges must be very sharp.

While the perceived definition of an image is determined primarily by the width of the edges between light and dark areas, there are second-order effects that depend on the *shape* of the edge transition. Compared with that of film, the slope of the transition curve for a television system is steeper in the center of the brightness scale, and there is often an overshoot. This tends to give the television image a "harder" appearance that is sometimes described pejoratively by the film industry as "the television look." The perceived definition of an image is also affected by the gray scale and the signal-to-noise ratio. These factors are treated in later sections.

The perception of an image's definition is closely related to its *limiting resolution* (see Sec. 2.6), which in turn is determined by the smallest angular separation at which individual lines can be distinguished. This ability is known as *visual acuity*. It can also be specified by the eye's *aperture response* (see Sec. 2.7).

The eye's visual acuity can be described by specifying the smallest angular separation at which individual lines can be distinguished. Visual acuity varies widely, as much as from 0.5' to 5' (minutes of arc), depending on the contrast ratio and the keenness of the individual's vision. An acuity of 1.7' is often assumed in the design of television systems.

The angular separation between lines can be converted to scanning lines after specifying the *viewing ratio*, the ratio of viewing distance to picture height, and this is a more common method of specifying this parameter.

With 440 active scanning lines (as in 525-line broadcast systems) and a visual acuity of 1.7', individual scanning lines can be distinguished at a viewing ratio as great as 4.4. At a viewing ratio of 4, the minimum usually assumed for broadcast systems, individual scanning lines can be faintly distinguished by a typical viewer.

A viewing ratio of 3 or less is usually assumed in designing HDTV, while a range of viewing ratios from 4 to 8 is assumed for broadcast television.

The difference in the eye's acuity for detail and edge sharpness at the viewing ratios assumed for HDTV and broadcast television systems establishes some of the most important differences in their design requirements. HDTV pictures, observed at a low viewing ratio, must have higher definition to give the same perception of sharpness. Broadcast-quality pictures seen from a low viewing ratio will appear fuzzy. On the other hand, an HDTV picture will appear no sharper than a broadcast-quality picture at greater viewing ratios. The practical result is that HDTV requires a larger screen at normal viewing distances to take advantage of its superior definition and signal-to-noise ratio. Conversely, a large screen image will appear "soft" and noisy with broadcast-quality definition and signal-to-noise ratio. See Chap. 15.

2.5.2 Criteria of image definition

Although the perceived definition of an image is determined by the sharpness of its edges, edge width is not a satisfactory working criterion for image definition. It is difficult to measure and equally difficult to handle analytically. As a result, television image definition is usually specified by indirect criteria, such as limiting resolution or aperture response. The measurement of aperture response is more complex, but it is a more complete criterion and is indispensable for system analysis. Limiting resolution is easy to measure and is particularly useful for routine maintenance.

2.6 Limiting Resolution

Limiting resolution has been used historically to specify the performance of optical instruments and film systems. It indicates the maximum number of alternate light and dark lines (or points of light) that can be distinguished per unit distance in the image. It was initially borrowed from film industry practices to specify the performance of television systems.

Its value as a criterion of the more basic property of image definition depends on the correlation between the number of lines that the system resolves and the sharpness of the edges in the picture. In film systems the correlation is generally good, and limiting resolution continues to be a standard criterion of definition in the film industry.

Limiting resolution has been less successful as a criterion for definition in television. Difficulties were encountered when it was used to compare the performance of television and film systems. Film systems with much higher limiting resolution often produced no sharper images than lower-resolution television systems. This is because the image definition is determined by its aperture response at all line numbers, whereas the limiting resolution is only one point on this curve, the maximum line number at which there is a discernible response. In spite of its flaws as a criterion of image definition, however, limiting resolution plays a useful role in the design and maintenance of television systems, and it continues to be widely used.

The Electronic Industries Association (EIA) publishes a standard resolution test chart, Fig. 2.2, that has been widely used for evaluating system performance. Limiting resolution is easy to measure, and the result is a single, easily understood number.

Because limiting resolution specifies only one point on the aperture response curve, it is unsatisfactory for comparing two systems if the shapes of their aperture response curves are greatly different. It is particularly unsatisfactory for comparing film and television systems, which have aperture response curves that are quite different.

2.6.1 Bandwidth and horizontal resolution

The horizontal limiting resolution of a television system—the number of black and white vertical lines that can be distinguished in a dimension equal to the picture height—is usually determined by its electrical bandwidth. In a conventional analog system, each cycle can reproduce a black-and-white line pair, i.e., two television lines. The horizontal resolution of analog systems as a function of bandwidth is given by Eq. (2.1), the inverse of Eq. (1.1). (This equation assumes that the aperture response of the system components is adequate to make the lines visible.)

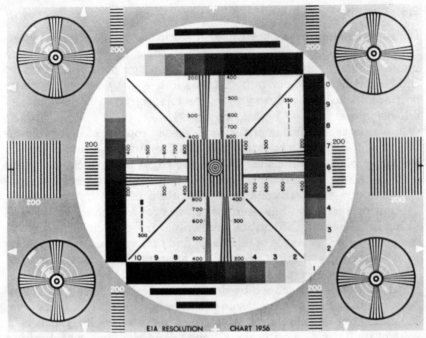

Figure 2.2   EIA resolution test chart. (*EIA.*)

$$R_H = \frac{2C_H B_w}{A_R N_L F_R} \tag{2.1}$$

where R_H = horizontal resolution

C_H = fraction of time in each scanning line devoted to the transmission of picture information after subtracting the time required for horizontal blanking

B_w = system bandwidth

A_R = aspect ratio

N_L = total number of scanning lines per frame

F_R = frame rate per second

The horizontal resolution R_H of NTSC broadcast systems (the United States and Japan), PAL broadcast systems (Europe), and the HDTV system developed by NHK in Japan (one of the first) are compared in Table 2.1.

The 20.0-MHz bandwidth used for the NHK HDTV system is impractical for present-day broadcast systems and difficult to achieve for cable and satellite. Much of the HDTV engineering research currently in progress is devoted to the development of analog and digital sys-

TABLE 2.1   Horizontal Resolution of Broadcast and HDTV Systems

	NTSC	PAL	HDTV
B_w (MHz)	4.2	5.0	20
N_L	525	625	1125
C_H	0.85	0.80	0.83
Aspect ratio	1.33 (4/3)	1.33 (4/3)	1.78 (16/9)
F_R (frames per second)	29.97	25	30
R_H (lines)	340	409	593

tems that will approximate the performance of conventional wideband analog systems but with a narrower bandwidth. See Chaps. 4 and 14.

2.6.2   Scanning lines and vertical resolution

The vertical limiting resolution of a television system—the number of horizontal lines that can be distinguished in a dimension equal to the picture height—is limited by the number of active (visible) scanning lines. It does not equal this number, however, because vertical detail in the image is randomly located with respect to the scanning lines. Consequently, the vertical resolution is significantly less than the number of scanning lines.

The ratio of the vertical resolution to the number of scanning lines is called the *Kell factor*. It is not a precise number because it is subjective and depends on the width of the scanning lines. It is commonly assigned a value of 0.7 on the basis of extensive observations, many of them made during the early years of television research.

The vertical resolution R_V is given by Eq. (2.2):

$$R_V = C_V K_f N_L \qquad (2.2)$$

where C_V = fraction of scanning lines that are visible (after subtracting the lines that are removed by vertical blanking)
   K_f = Kell factor
   N_L = total number of scanning lines

The vertical resolutions of NTSC broadcast, PAL broadcast, and a typical HDTV system are compared in Table 2.2.

TABLE 2.2   Vertical Resolution of Broadcast and HDTV Systems

	NTSC	PAL	HDTV
C_V	0.92	0.92	0.96
N_L	525	625	1125
K_f	0.7	0.7	0.7
R_V (lines)	343	408	767

The vertical resolution of the HDTV system exceeds the horizontal in order to minimize aliasing, a beat pattern between the scanning line pattern and vertical detail in the picture; see Sec. 2.17.6.

2.6.3 Picture elements (pixels)

A television raster can be described as a grid of picture elements, or pixels. The height of each pixel is the spacing between scanning lines divided by the Kell factor, and its width is the space on a scanning line occupied by one-half cycle of the highest frequency in the transmission bandwidth. The number of pixels, which is equal to the product of the bandwidth and the fraction of time occupied by visible scanning lines, is a rough indicator of the information content of the image.

The number of pixels is:

$$N_P = \frac{2CK_fB_w}{R_F} \tag{2.3}$$

where C = fraction of time occupied by active scanning (after subtracting vertical and horizontal blanking periods)
K_f = Kell factor
B_w = bandwidth of the system, hertz
R_F = frame rate

In the broadcast and HDTV systems described in the preceding sections, the broadcast system has 148,000 pixels, while the HDTV system has 743,000.

2.7 Aperture Response

Aperture response is a universal criterion for specifying picture definition and other aspects of imaging system performance. It can be used for film images, camera lenses, television camera imagers, video amplifiers and other bandwidth-limiting components, the scanning process, receiver picture tubes, and the human eye. It is used frequently in this volume to describe the performance of television system components.

2.7.1 Definition of aperture response

The term *aperture response* is defined in Fig. 2.3. Assume that a pattern of black and white lines of varying widths is scanned by a narrow light beam (the aperture), and the peak-to-peak variation in the reflected light from the black and light lines is measured. On lines that are much wider than the diameter of the spot, these variations will be

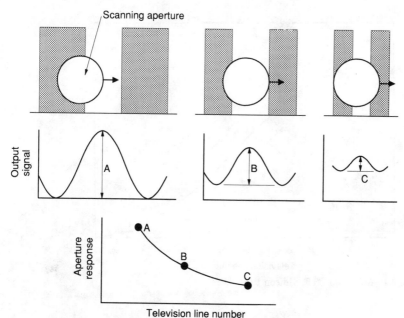

Figure 2.3   Definition of aperture response.

of full amplitude. As the width of the lines is decreased so that the scanning spot always overlaps a portion of both a black and a white line, the amplitude of the variations will decrease. When the width of the lines is twice the diameter of the spot, the variations disappear.

The width of these lines is specified by its reciprocal, the number of alternate black and white lines (counting both black and white lines) that can be fitted into the vertical dimension of the picture. This parameter is known as the *television line number*.

The aperture response of a component or system is a graph of the peak-to-peak amplitude of its response (in the case of Fig. 2.2, of the variations in reflected light) as a function of the television line number.

The aperture response of a component or system can be specified either by its response to a square-wave pattern, i.e., alternate dark and light bars, known as the contrast transfer function (CTF), or by its response to a theoretical pattern in which the cross-sectional darkness of the bars varies sinusoidally, the modulation transfer function (MTF). The CTF is physically measurable, but the MTF is more useful for analytic purposes. The conversion can be made by means of Eq. (2.4).

$$MTF_n = \left(\frac{\pi}{4}\right)\left(CTF_n + \frac{CTF_{3n}}{3} - \frac{CTF_{5n}}{5} + \frac{CTF_{7n}}{7}\cdots\right) \qquad (2.4)$$

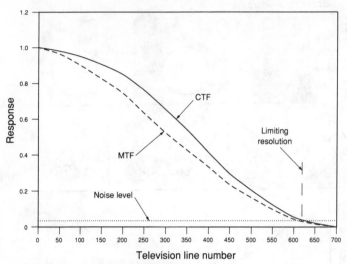

Figure 2.4   MTF and CTF, vidicon tube.

where MTF_n = modulation transfer function (aperture response to a sinusoidal pattern) at line number n

CTF_n = contrast transfer function (aperture response to a square-wave pattern) at line number n

The aperture response curves, MTF_n and CTF_n, for a vidicon pickup tube, which has a response that is typical of television system components, are plotted in Fig. 2.4.

2.7.2   Aperture response of components in series

The aperture response of a system that has several components in series can be calculated by multiplying the MTFs of the individual components, just as with the frequency response of cascaded electrical components. The ability to calculate the cascaded aperture response of a number of components in this manner is a very useful characteristic of aperture response as a criterion of performance.

2.7.3   Limiting resolution and aperture response

The line number at which the aperture response curve disappears into the noise is known as the *limiting resolution* (see Fig. 2.2). Although this is not as comprehensive a criterion of picture quality as image definition, it is frequently used because of its simplicity and the ease with which it can be measured. Also, it is a customary method of measuring the performance of film and other components of photographic systems.

2.7.4 Aperture response of the human eye

Figure 2.5 shows the aperture response, MTF, of a typical human eye at viewing ratios of 3, 4, and 8.[1] This response function should be considered as nominal rather than precise because the visual acuity of individuals varies widely.

2.7.5 Aperture response as a measure of image definition

The correlation of aperture response with definition or edge sharpness is not intuitively obvious, but it can be demonstrated by calculating the output of a system of known aperture response with a step function input, i.e., an input signal that makes an infinitely steep transition between two levels. The sharpness of this transition as a function of the system aperture response is calculated by an *inverse Fourier transform*.

$$e(x) = \int_{-\infty}^{\infty} g(n)e^{j\omega n}\, dx \qquad (2.5)$$

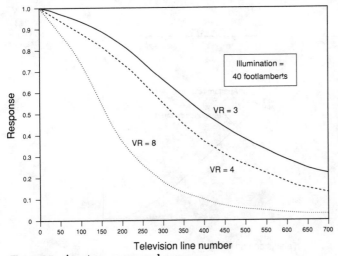

Figure 2.5 Aperture response, human eye.

where $e(x)$ = amplitude response of the component or system as a function of its position on the scanning line, i.e., along the x axis of the raster

$g(n)$ = aperture response of the system as a function of the television line number $n = 2\omega x$

With the aid of this equation, the correlation between aperture response and definition can be verified.

2.7.6 Equivalent line number N_e

An aperture response curve is the most complete indication of the definition of a system, but it is not as convenient or as easily compared as a single number. In an effort to define a single number that would specify the definition of a system to a reasonable degree of approximation, Schade[2] developed the concept of equivalent line number, or N_e. Although N_e is not widely used as a criterion of picture quality, it is very useful for system analysis.

N_e is defined in Fig. 2.6. It is the line number that defines a rectangle having the same area as the area under the aperture response squared curve. (In the electrical analog, it is the *noise bandwidth*.)

The validity of N_e as a measure of picture definition can be demonstrated by calculating the edge configurations of systems having the same N_e but differently shaped response curves. Figures 2.7 and 2.8 compare two imaging systems with the same N_e but with aperture response curves typical of television and film system components. Figure 2.7 shows their aperture response, while Fig. 2.8 shows their re-

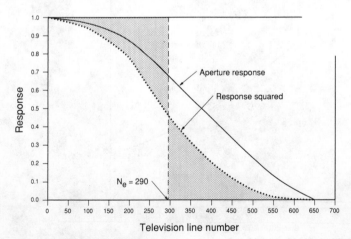

Figure 2.6 Definition of N_e.

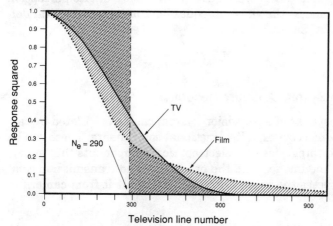
Figure 2.7   Aperture response—film and TV.

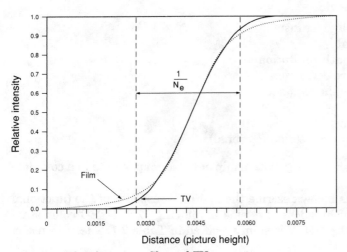
Figure 2.8   Edge sharpness—film and TV.

sponse to a step function input. Note that the resolution of the film system is much higher, but its response is lower at lower line numbers. This causes minor differences in the shape of the edges, but the edge width is approximately the same, confirming the validity of N_e as a measure of image definition. The edge width is given by

$$\text{Edge width} = \frac{\text{picture height}}{N_e} \tag{2.6}$$

Within reasonable limits of inequality, the approximate perception of definition of images that have unequal horizontal and vertical definition is given by Eq. (2.7).

$$N_e = (N_{e_{vert}} N_{e_{horiz}})^{1/2} \qquad (2.7)$$

2.8 Television System Aperture Response

The aperture response of a television system can be calculated by cascading the response curves of its functional components. These can be divided into two categories: the electrooptical components that generate and display the images, and the processing and transmission components that process the video signal and transmit it from camera to receiver:

Electrooptical Components
   Camera lens
   Camera pickup device (imager)
   Display device
Processing and Transmission
   Horizontal
      Bandwidth limitation
      Aperture correction
      Image enhancement
   Vertical
      Scanning
      Aperture correction (optional)

Figure 2.9 shows the relationship of these components in a complete system.

The sections below describe the aperture response of the functional components of broadcast systems employing U.S. NTSC color standards and of the HDTV system specified in Table 2.1. The response of

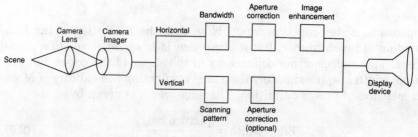

Figure 2.9 Aperture response determining components.

the video transmission channel and the scanning pattern would be somewhat different for a system using PAL standards.

2.8.1 Electrooptical components— broadcast systems

Aperture response curves for typical broadcast-quality electrooptical components and their overall response when cascaded are shown in Fig. 2.10. The performance of commercial products varies widely, and these curves are representative and not intended to describe any specific models.

2.8.2 Camera lens

A high-quality, fixed-focus lens such as that used as an example in Fig. 2.10 has a very high response out to 300 lines, the approximate upper limit of NTSC systems, and has only a minor effect on the performance of these systems. Zoom lenses make some compromises in aperture response in order to provide the variable-focus feature, and the performance of these lenses must be considered when they are used in HDTV systems.

2.8.3 Camera imager

The camera imager aperture response curve shown in Fig. 2.10 (and also in Fig. 2.4) is for a vidicon tube; its response, which is measured

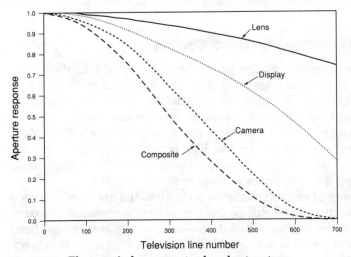

Figure 2.10   Electrooptical components—broadcast systems.

by the peak-to-peak amplitude of its output signal, is typical of the camera transducers used in the U.S. broadcast industry.

2.8.4 Display device

The display device aperture response curve shown in Fig. 2.10 is for a shadow-mask color kinescope (see Chap. 3) with 715 vertical slot apertures per picture width or a television line number of 535 (715/1.33). Aliasing, in this case the interference pattern between the slot apertures and horizontal detail, begins to appear at about 535 lines. The horizontal resolution, before it is limited by other factors, is 400 lines (535 lines times a Kell factor of 0.7). The limiting resolution of the kinescope is above the range of the video passband in U.S. broadcast systems, but aliasing might occur if the amplitude of the high-frequency components in the signal were sufficient.

2.8.5 Cumulative response, electrooptical components

The cumulative aperture response of the electrooptical components in the example above is approximately 0.4 at a television line number of 340, the upper limit established for the U.S. broadcast system by the video passband and the number of scanning lines. The response can be improved further by the use of aperture correction (see below) to give better results within the bandwidth and scanning line limitations of the system. The composite response of broadcast-quality components is marginal or unsatisfactory for HDTV systems. This dictates use of higher-definition camera transducers and display devices.

2.9 Vertical Definition—Broadcast Systems

The aperture response of the scanning pattern, which affects the sharpness of horizontal edges and the reproduction of vertical detail, is not a straightforward single-valued function, but is the complex result of the size of the scanning aperture and the sampling process.

The sampling process in itself does not assure the aperture response out to the limit established by the Kell factor because the scanning devices, the camera imager, and the display device have finite apertures with aperture responses such as shown in Fig. 2.10.

In the absence of vertical aperture correction, the aperture response in the vertical dimension will be the cumulative response of the components out to the limiting resolution—the number of scanning lines multiplied by the Kell factor. For NTSC broadcast systems, this is approximately 340 lines (see Table 2.2). Detail is not satisfactorily reproduced above this limit.

Components of vertical detail above the number of active scanning lines, or about 485 lines in a 525-line system, will produce aliasing. Aliasing occurs in fine vertical detail as a result of a beat between line number components greater than the number of active scanning lines. Its severity is determined by the magnitude of the signal energy at television line numbers in the vicinity of the number of scanning lines. Figure 2.10 shows that the cumulative response of the electro-optical components (including the receiver picture tube) for broadcast systems is so low at these line numbers that aliasing is only a minor problem. With HDTV systems having electrooptical components with higher responses at these line numbers, this defect can be serious, and the choice of the number of scanning lines may be based as much on avoiding aliasing as on increasing resolution.

Vertical aperture correction circuits have been developed using delay lines that make it possible to compare the video signals on adjacent lines. The aperture corrector accentuates differences in signal levels between corresponding points on adjacent lines as would occur on a horizontal edge, thus sharpening its appearance. (The same technique can be used for horizontal image enhancement by the use of delay lines that delay the signal by one picture element.)

2.10 Horizontal Definition—
Broadcast Systems

The horizontal definition of a broadcast channel is determined by the cascaded aperture response of the electrical channel and the electro-optical components. Figure 2.11 shows three channel responses: (1) with flat amplifier response to the band limit of 4.2 MHz (U.S. NTSC); (2) with *aperture correction* in the video passband to compensate for the fall-off in component response; and (3) with aperture correction and *image enhancement* to enhance the image sharpness further.

To calculate the response of the aperture correction and image enhancement circuits, the television line numbers are translated to electrical frequencies by Eq. (2.8):

$$f = \frac{(\text{TV line number})A_R N_L R_F}{2C_H} \tag{2.8}$$

where C_H = fraction of time in each scanning line devoted to the transmission of picture information (after subtracting the time required for horizontal blanking)

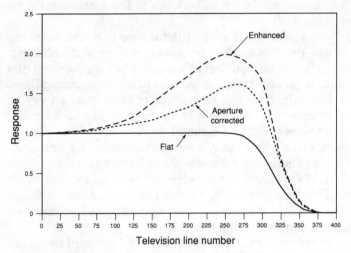

Figure 2.11   Horizontal aperture response—broadcast systems.

f = frequency
A_R = aspect ratio
N_L = total number of scanning lines per frame
R_F = frame rate per second

The added gain for aperture correction is usually adjusted so that the combined response of the electrooptical components and the electric circuits remains constant out to about 400 television lines.

Image enhancement is achieved by further increasing the gain of the video channel at the frequencies corresponding to 200 to 300 lines. This increases the sharpness of the picture, but the amount of enhancement must be limited to avoid undesirable "ringing" and unnatural edge effects.

The response of an electrical system to a video signal is determined not only by its amplitude/frequency response, i.e., its aperture response and image enhancement, but also by its phase/frequency response. This must be carefully controlled so that the leading and trailing edges of transitions are symmetrical.

Aperture correction and image enhancement can be introduced at the camera or at the receiver (or both). Introducing them at the camera results in a smaller increase in noise, while introducing them at the receiver makes it possible to adjust the degree of enhancement while observing the end result.

2.11 N_e of Broadcast TV Systems

The horizontal N_e of the system described in Fig. 2.10 with aperture correction is about 310; this is good performance for a broadcast system. The performance of the image-enhanced system is even better, with an N_e of 400, but the high N_e combined with the sharp frequency cutoff requires very careful control of the phase/frequency response to minimize overshoots on the picture edges. This is very nearly the practical limit that can be achieved in a broadcast system without causing an unnatural oversharp appearance of the image edges.

The system with the vertical response shown in Fig. 2.10 does not have aperture correction, and it has an N_e of 185. Adding vertical aperture correction would both improve the sharpness of the picture and more nearly balance the horizontal and vertical definitions. Note that major aliasing begins at a television line number where the response of the electrooptical components is low, and the resulting moiré pattern would be barely visible. Adding vertical aperture correction, however, might result in visible moiré.

2.12   HDTV System Aperture Response

The determination of the aperture response is not as straightforward for many of the proposed HDTV systems as for NTSC and PAL systems. A number of them employ digital compression and other bandwidth-reduction techniques to reduce the bandwidth requirements while retaining the perceived definition of conventional widerband analog systems. One technique, for example, is to transmit the stationary portions of the image in high definition but at a slower rate while transmitting the moving portions with reduced bandwidth. This method takes advantage of the eye's reduced acuity to detail for moving objects. Other systems are described in Chaps. 4 and 14.

For the purpose of comparing the performance of typical NTSC and HDTV systems, the NHK MUSE-E HDTV system specified in Tables 2.1 and 2.2 is used as an example. It achieves its performance primarily by the use of a 20-MHz bandwidth rather than by special bandwidth-reduction schemes. Thus, the same methods of analysis can be used to compare its performance with that of conventional NTSC systems.

2.12.1   Electrooptical components

The higher definition of HDTV systems cannot be achieved solely by increasing the bandwidth and number of scanning lines. High-definition lenses, camera imagers, and display devices are also required.

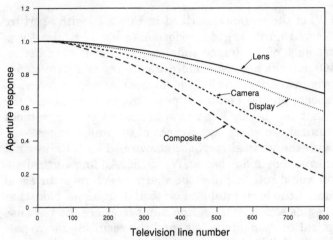

Figure 2.12   Electrooptical components—HDTV systems.

Figure 2.12 shows representative aperture responses for electrooptical components that are required to achieve HDTV performance. The N_e of the cascaded components is 400, compared with 240 for broadcast-quality components.

2.12.2   Vertical definition

In the absence of vertical aperture correction, the vertical aperture response would be that of the electrooptical components out to the limiting resolution of about 700 lines. The vertical N_e of this system would be about 400 lines. The high aperture response of the components at 700 lines and above indicates the importance of increasing the number of scanning lines beyond that established by the limiting resolution requirement to reduce or limit aliasing.

It is expected that the use of vertical aperture correction will become common in future HDTV systems as the result of technical developments that make it practical:

1. The large number of scanning lines minimizes aliasing.

2. Inexpensive memory circuits, a prerequisite for vertical aperture correction, are now available.

3. Higher-quality electrooptical components are now available that limit the amount of aperture correction (and the attendant loss of signal-to-noise ratio) that is necessary.

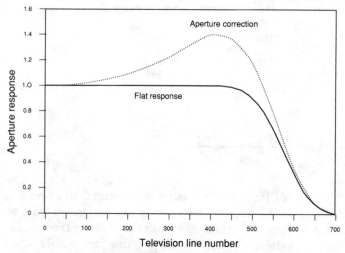

Figure 2.13   Horizontal aperture response—HDTV systems.

2.12.3   Horizontal definition

Figure 2.13 shows the aperture response of an electrical channel for a 1125-line system with and without aperture correction. As with the broadcast channel, the horizontal definition is determined by the cascaded response of the channel and the components. Image enhancement may not be necessary or desirable, since the image would be quite sharp without it and enhancement affects the signal-to-noise ratio and aliasing.

2.13   Comparison of Film and Television Definition

It was stated earlier that an objective in the design of broadcast systems is to match the definition of 16-mm film. Similarly, it is hoped to match the definition of 35-mm theatrical film with HDTV.

Table 2.3 indicates the success of this effort by comparing the limiting resolution and N_e of representative broadcast (NTSC) and HDTV television systems (no vertical aperture correction for NTSC) with those of projected film images. In each case, the TV system performance is the cascaded component performance, Fig. 2.10 or 2.12, and the channel response, Fig. 2.11 or 2.13.

This table demonstrates the inadequacy of limiting resolution as a criterion for comparing the image definition of different types of sys-

TABLE 2.3   Comparison of Film and Television Performance

	Limiting resolution	N_e
16-mm film	1100	300
NTSC TV		
Horizontal	312	310
Vertical	312	310
35-mm film	1800	450
HDTV		
Horizontal	553	513
Vertical	767	481

tems. The resolution of film is higher, but the perceived sharpness is about equal because of the greater response of television systems at lower line numbers. This is fortunate, because matching the resolution of film in a television system would require impossibly wide bandwidths.*

In practice, the N_e of both film and television systems varies widely, and a range of performances should be compared, as shown in Table 2.4.

TABLE 2.4   Ranges of Film and Television Performance

	N_e
16-mm film	200–300
NTSC TV	175–310
35-mm film	350–450
HDTV	300–513

The table shows that television systems have the capability of producing image definition that is equal to that of film. With more variables to introduce degradation, however, the actual performance of television systems is often less.

It should also be noted that the appearance of TV and film images will never be exactly the same because of the differing shapes of their aperture response curves, as shown in Fig. 2.14.

---

*Schade's work[2] was the classic study of this comparison.

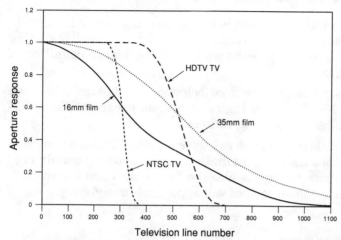

Figure 2.14   Comparison of television and film aperture response.

2.14   Visual Perception of Broadcast and HDTV Images

Figure 2.15 shows the combined aperture responses of the eye and of NTSC and HDTV television systems at viewing ratios of 4 and 8. They are calculated by cascading the response of a normal eye as shown in Fig. 2.5 with that of NTSC and HDTV systems as shown in Figs. 2.11 and 2.13.

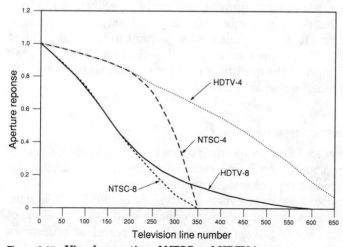

Figure 2.15   Visual perception of NTSC and HDTV images.

Figure 2.15 illustrates the following:

1. At a viewing ratio of 8, there is very little difference in the perception or appearance of NTSC and HDTV systems. A smaller viewing ratio is necessary to take advantage of the higher definition of HDTV images.

2. As the viewing ratio is reduced below 8, the difference between HDTV and NTSC images begins to become apparent. The difference is striking at a viewing ratio of 4.

3. At viewing ratios of 8 and above, the eye rather than the television system becomes the major limitation in the perceived definition of the image. Since the eye is also the limitation when the scene is viewed directly, these images will appear relatively sharp.

2.15   Image Gray Scale

The importance of gray scale as a measure of image quality can be demonstrated by turning on the room lights while photographic slides are being projected on a screen. With the lights out, the images appear crisp and in focus. When the room lights are turned on, increasing the ambient lighting on the screen, the images appear washed-out and fuzzy. The addition of a fixed amount of ambient illumination to the slide image over a large area of the screen distorts not only the image's gray scale but also its apparent definition.

In color systems, as described in Chap. 3, careful control of the *transfer function* (the parameter that specifies gray scale; see below) of the three primary color signals is required to avoid color distortion.

Unlike image definition, improved gray scale reproduction requires no changes in bandwidth, number of scanning lines, or other transmission standards. It is nevertheless as important to the production of high-quality HDTV pictures as improved definition. A better gray scale will also improve the quality of standard broadcast pictures with no change in transmission standards.

2.15.1   Gray scale transfer function

The gray scale of an image is specified by its *transfer function*, a plot of image brightness as a function of scene brightness. Figure 2.16 is the transfer function of a typical system, including camera, video channel, and display device. It is the television equivalent of the H & D (Hurter and Driffield) curves that specify film density as a function of exposure for photographic negatives. Because the eye perceives brightness

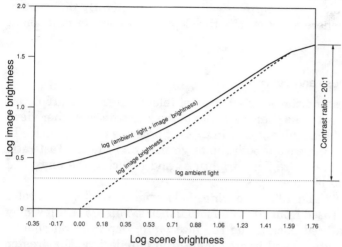

Figure 2.16   Television transfer function.

differences in terms of ratios rather than numerical differences, a logarithmic scale is used for both scene and image brightness.

The transfer function can also be used to specify the input-output performance of television system components, including camera imagers, video amplifiers, and display devices. The system gray scale function is obtained by cascading the transfer functions of its components.

Like that of most imaging systems, the television transfer function is an S-shaped curve with a linear section in the middle that slopes off at both ends. The upper limit of the curve is called the *white field brightness* and is usually established by the limitations of the display device, ordinarily a kinescope. The lower limit is called the *unexcited field brightness* and is the brightness of the kinescope when it is turned off; this is primarily the result of reflection of ambient light.

It is often convenient to specify the gray scale of a television image by numerical parameters rather than graphically. The three most commonly used parameters, as shown in Fig. 2.16, are as follows:

1. *Highlight brightness:*   The white field brightness, i.e., the brightness of the brightest areas of the image.

2. *Contrast ratio:*   The brightness ratio of the brightest and darkest areas of the image. For an imaging system, it is the ratio of the white field and unexcited field brightnesses.

3. *Gamma:*   The slope of the transfer function curve. Gamma is a term borrowed from photography, where it is used to specify the

slope of the H & D curve at the single point in its linear portion where the slope is the greatest. Its usage is slightly different in television, where it can specify the slope of the transfer function at any point.

2.15.2 Brightness and contrast ratio

The brightness and the contrast ratio of television images are determined primarily by the characteristics of the kinescope (or other viewing device) and the ambient lighting. Great progress has been made in recent years in increasing both the brightness and the contrast ratio by increasing the white field brightness and reducing the unexcited field brightness of kinescopes.

Between 1962 and 1982, the white field brightness of typical color tubes was increased from 10 to 65-foot lamberts. This was achieved by utilizing metallized backing on the face plate, improved phosphors, higher beam voltage, and improved electron gun design. The greater brightness of the phosphors was achieved by deviating somewhat from the NTSC primary color specifications (see Chap. 3). This has not been a serious concern for broadcast systems, but it has been questioned for HDTV systems that strive for perfection.

Most of the unexcited or dark field brightness results from reflection of ambient light from the picture tube face. Its effect was originally reduced by placing a neutral-density filter in front of the kinescope. Light from the kinescope passed through the filter once, while light reflected from the kinescope passed through it twice. If the filter transmitted one-half the light, the white field brightness would be reduced by one-half, but the unexcited field brightness would be reduced by three-quarters and the contrast ratio doubled.

The unexcited field brightness of modern color kinescopes has been reduced by the *black matrix/black surround* and other techniques that substantially reduce the light reflected by the phosphors. The results can be seen by observing the kinescope of a color receiver of recent design when the set is turned off. In contrast with older sets in which the kinescope was whitish gray, new sets appear black.

A high contrast ratio is a technical ideal and, with some exceptions, an aesthetic one as well. Improved kinescopes have made it possible to meet the desired contrast ratio standards for high-quality broadcast TV, provided the ambient lighting is not excessive. Optimum results for HDTV require control of ambient lighting by creating a theater-like environment for home viewing.

Because of the eye's logarithmic response to brightness differences, the reflected ambient light from the kinescope screen, which establishes the unexcited field brightness, can have a greater effect on the

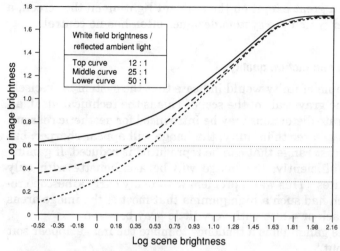

Figure 2.17   Effect of ambient light.

contrast ratio than the highlight or white field brightness. A relatively small numerical difference in reflected light can have a major effect on the contrast ratio.

This effect is illustrated in Fig. 2.17, which shows the transfer function under three different conditions. All assume a bright field brightness of 50 foot-lamberts, but with unexcited field brightnesses of 1, 2, and 4 foot-lamberts as a result of differences in ambient lighting. The resulting contrast ratios are 50, 25, and 12.5. The gamma is also reduced as the ambient light is increased. The gamma of the image with low ambient lighting is 0.94, and it decreases to 0.76 as the lighting is increased.

The combination of a lower contrast ratio and a reduced gamma with high reflected light results in a softer, less sharp picture, the opposite of the effect desired for HDTV and undesirable for broadcast television.

Table 2.5 shows the approximate ranges of contrast ratios that are

TABLE 2.5   System Brightness and Contrast Ratio

System	Brightness (foot-lamberts)	Contrast ratio
Motion picture theater	25	50–100
HDTV (desired)	100	50–100
Broadcast TV (high-quality)	65	30–65
Broadcast TV (typical)	40	10–20

typical of or desired for operating systems. In practice, the contrast ratio depends to a great extent on the ambient lighting on the screen, a parameter over which the system designer often has no control.

2.15.3 Gamma and picture quality

An image gamma of unity would indicate that the system is precisely reproducing the gray scale of the scene. This is the technical ideal, although deliberate distortions may be introduced for aesthetic reasons.

If gamma is greater than unity, the image will appear sharper, but the scene contrast range that can be reproduced is reduced. If gamma is increased sufficiently, the image will be a silhouette with only blacks and whites. (This was a problem with many early kinescope recordings, which had such a high gamma that most of the image areas were either black or white with very little gray between.)

Reducing the gamma has a tendency to make the image appear soft and "washed out."

Determining the gamma of a television picture that is most pleasing is an aesthetic judgment. For color images, a gamma of 1.2 to 1.5 is generally accepted as optimum.[3]

2.15.4 Gammas of television images and components

The gamma of the image produced by a complete television system can be calculated by multiplying the gammas of its individual series components, specifically the camera imager, the display device, and any nonlinear video amplifiers that are added to the system for *gamma correction.*

With the exception of vidicons, which have a gamma of about 0.7, most camera imagers have a gamma of unity.

Kinescopes, which are almost universally used as display devices, have gammas from 2.2 to 2.4.

Gamma correction circuits usually have a gamma of less than unity to compensate for the high gammas of kinescopes. For example, if it is desired to have a system gamma of 1.2 with a unity-gamma camera imager and a kinescope gamma of 2.2, the gamma correction circuit should have a gamma of 1.2/2.2, or 0.55.

2.15.5 Gamma correction and system noise

Gamma correction circuits with a gamma of less than unity "stretch" the black portion of the signal, which causes noise in the black areas of the picture to be amplified more than noise in the white. The effect on the perceived signal-to-noise ratio in the image will depend on the location of the gamma correction circuit with respect to the noise source. Noise generated in the dark areas of the picture ahead of the gamma corrector, e.g., in the imager preamplifier, will be amplified more than the picture signal. Depending on the picture content, this may reduce the overall signal-to-noise ratio by several decibels. The adverse affect of this phenomenon is exacerbated by the fact that noise is more visible in the dark areas.

Conversely, dark-area noise generated after gamma correction, as in the transmission link, will be amplified no more than the signal, and its amplitude will be reduced by the high gamma of the kinescope.

2.16 Signal-to-Noise Ratio

2.16.1 Sources and characteristics of noise in television systems

Random electrical disturbances, which are usually thermal in origin, produce "snow" in the picture and are the most common type of noise in video signals. The relationship between temperature and noise is close, and *noise temperature* (see Sec. 11.6.5) is a frequently used specification for system components.

The most significant sources of random noise are components where low signal levels are encountered. They include camera imagers, camera preamplifiers, videotape recorders, cable circuits, microwave and satellite circuits, and broadcast receivers. The noise generated by these elements establishes basic limitations on system performance and is fundamental in establishing design parameters.

Most random noise is "white," i.e., the noise energy is uniformly distributed through the frequency spectrum, as with white light. Important exceptions are the noise generated in microwave and satellite links and by photoconductive imagers. These system components have the characteristic that the noise power generated per hertz increases as the square of the frequency. The links display this effect because of their use of frequency modulation (see Sec. 11.9.2). The noise generated by photoconductive transducers results from the characteristics of their output circuits (see Chap. 5).

2.16.2 Definition of signal-to-noise ratio

The signal-to-noise ratio is the basic specification for random noise. It is the power ratio, usually expressed in decibels, of the peak-to-peak signal voltage or current, from black level to reference white but not including the sync pulses, to the rms value of the noise.

2.16.3 Unweighted and weighted noise

Noise power can be specified as either *unweighted* or *weighted*. Unweighted noise is measured by an instrument with uniform frequency response. Weighted noise, a measure of noise visibility, is measured with an instrument having a frequency response, the weighting function, that simulates the cascaded aperture response of the eye (see Fig. 2.5) and the system components following the noise measurement point.

The variables in the calculation of the weighting function include the viewing ratio, the assumed response of the eye, the screen brightness, the width of the scanning lines, and the aperture response of the kinescope and other system components following the point of measurement. There are differences between the visibility of noise in monochrome and color images because of interaction of the high-frequency color subcarrier and noise components in this frequency range. The weighting factor, therefore, cannot be both universal and precise.

The industry has found it important, however, to establish a standard weighting function that approximates practical situations so that systems can be specified and compared on a uniform basis. The Electronic Industries Association has published a widely used standard for broadcast transmission systems, which is part of EIA/TIA-250-C and is shown in Fig. 2.18. It is based on a somewhat more pessimistic assumption as to the aperture response of the eye and system components than is indicated by Figs. 2.5 and 2.11.

The difference between the weighted and unweighted signal-to-noise ratios is called the *weighting factor*. Using the weighting func-

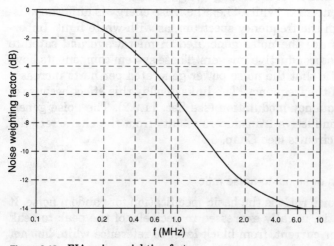

Figure 2.18   EIA noise weighting factor.

tion shown in Fig. 2.18, it is about 8 dB for white noise and 12 dB for FM noise. It is larger for FM noise because most of the noise is concentrated at the upper end of the transmission spectrum, where its visibility is less and the weighting function is lowest.

A standard industry weighting factor for HDTV does not presently exist, but it has been variously estimated as being 3 to 5 dB *less* than for broadcast service. This would indicate that the unweighted signal-to-noise ratio for HDTV should be 3 to 5 dB higher for the same weighted value. The weighting factor is smaller because a lower viewing ratio is assumed and fine-grained high-frequency noise is more visible.

2.16.4 Signal-to-noise ratio in digital systems

Noise power is additive in analog systems, and the signal-to-noise ratio for the complete system is inevitably lower than that of the worst stage. Similarly, with multiple-generation recordings, the signal-to-noise ratio of the final generation is lower than that of any individual generation.

With digital transmission or recording (see Chap. 4), noise causes *bit errors*, but so long as the *bit error rate* (ber) is low enough, correction circuits can restore the original bit stream and there is virtually no degradation of the signal-to-noise ratio. If the noise power is increased sufficiently, however, the bit error rate becomes so large that error correction is not effective, and the signal is lost.

In summary, an analog signal suffers *graceful degradation* as it passes through successive stages or recording generations. Its quality gets steadily worse, but it remains useful as noise is added. By contrast, the signal-to-noise ratio in digital systems suffers little reduction as the noise increases to a critical level, but at that level the signal suddenly becomes unusable (also see Sec. 4.3.2).

2.16.5 Signal-to-noise ratio standards

A number of industry and governmental standards have been established for the video signal-to-noise ratio. Some are for a complete system from camera to receiver, while others cover only a portion of the system, e.g., a microwave link. Some are for white noise and others for FM noise. Some are for weighted and others for unweighted noise. When using published standards, it is important to determine exactly what is being specified.

Table 2.6 shows one commonly accepted set of signal-to-noise standards for broadcast systems. It also shows possible goals for HDTV system standards. The HDTV goals are very demanding, and possibly

TABLE 2.6 Signal-to-Noise Ratio Standards

Service	Weighted, dB	Unweighted (White), dB	Unweighted (FM), dB
HDTV	55	50	47
Broadcast			
Good	53	45	42
Fair	43	35	32
Poor	37	29	26

not always obtainable, but they would result in superb, noise-free pictures. The broadcast standards are based on a study by TASO[4] (Television Allocations Study Organization) for the FCC. Although the report is a number of years old, its conclusions are still reasonably valid.

Since noise is additive for analog transmission, a practical standard must take the number of system stages into account. As an example, the EIA *weighted* signal-to-noise standards for transmission circuits are as follows:

Circuit	S/N (dB)
Short haul	67
Medium haul	60
Satellite	56
Long haul	54
End-to-end system	54

2.17 Image Defects

The remainder of this chapter expands the description of the television image defects that were defined in Sec. 2.3. The magnitude of these defects, together with the system's performance with respect to the basic criteria of definition, gray scale reproduction, and signal-to-noise ratio, determines the quality of the images that are produced.

2.17.1 Hum

Hum bars were a major problem in the design of early television receivers. To alleviate the problem, field rates equal to this frequency (60 Hz in the United States, 50 Hz in Europe) were chosen. Spurious patterns resulting from hum were then stationary and less noticeable. Small deviations from this frequency became necessary as the industry developed. Networking brought signals into areas where the frequency of the power source was not necessarily synchronized with that at the program source. And NTSC color standards specified the use of a field rate (59.94 Hz, a submultiple of the subcarrier frequency) that was slightly below the nominal 60-Hz power source frequency. The effect of these small deviations, which causes the hum pattern to move slowly up or down on the raster, was eliminated by better receiver designs that greatly reduced the crosstalk of the power source into the video signal. The design of modern receivers has now been improved to the extent that hum should not be a problem even with large differences between power source and field frequencies.

2.17.2 Co-channel interference

The FCC originally specified a desired-to-undesired carrier ratio of 20: 1, 26 dB, or greater to reduce co-channel interference to an unobjectionable level. Subsequently it was discovered that the visibility of the interference (venetian blinds) could be greatly reduced by an offset carrier technique in which the carrier frequencies of adjacent co-channel stations were offset by ±1/2 the line rate, or ±7625 Hz, for a total difference of 15,250 Hz. The interference bars then have a width equal to that of the scanning lines and are less visible by about 20 dB than the wide bars that are produced when the carriers are only slightly separated in frequency.

This offset was applicable to two stations, but if a third station was added, two of the three would be on the same frequency. A compromise was adopted in which the carriers are separated by ±10 kHz. This results in an improvement of about 12 dB in the signal-to-noise ratio, and offset carriers are now specified by the FCC.

2.17.3 Lag

Lag usually originates in the camera imager and is caused by the failure of the scanning process to completely eliminate the electrical pattern on the imager's photosensitive surface in a single scan. As a result, motion in the image, whether caused by movement in the scene or movement of the camera, leaves a trail.

Lag was a relatively unimportant problem with image orthicons, but it is a serious problem with vidicons. Unfortunately, the operating mode that gives the vidicon maximum sensitivity (high target voltage; see Chap. 5) also produces maximum lag. As a result, vidicons have limited use for live pickup; they are used mainly in closed-circuit applications where image quality is not critical. For the pickup of film programs, however, sufficient light is available to reduce the lag of vidicons to an acceptable level, and they have been widely used for that purpose.

The Plumbicon and other more recent tubes have a much lower lag than vidicons at low light levels, and they are standard imagers for studio and field cameras. Even the Plumbicon, however, has a residual signal of about 3 percent on the third field following the removal of the image from the photosensitive surface.

Solid-state CCD (charge-coupled device) imagers are practically free of lag and are finding increasing use for live cameras. They received a dramatic introduction in the 1984 World Series when the stitches on a rapidly spinning pitched baseball were made visible.

Blooming and *comet tails* are defects related to lag in that they are spurious signals that result from the transducer's response to changes in light level. They are caused by overloading of the photosensitive surface with extremely bright highlights. They are described further in Chap. 5.

A related defect is *smear*, which is caused by inadequate frequency or phase response of the video channel. It is an extreme example of low definition (see above); the edges are widened, and because of poor phase response, the trailing edges of pulses may differ from their leading edges.

2.17.4 Geometric distortion

Geometric distortion of the final image may originate in either the camera or the display device. The distortion produced by the display device is usually much greater, in part because it is a mass-produced product with wider tolerances and in part because the very wide deflection angle required by most picture tubes makes scanning linearity more difficult.

The geometric distortion produced by a camera chain can be measured by the EIA linearity test chart shown in Fig. 2.19, sometimes called a ball chart.

The inner diameter of each circular ring is 1 percent of picture height, while the outer diameter is 2 percent. Linearity is measured by superimposing the image of this chart as produced by the camera on an electronically generated grating pattern in which the intersections of the vertical and horizontal lines should coincide with the centers of the circles. A well-designed camera will maintain the intersections within the 1 percent circles (±0.5 percent) in a circle in the center of the raster having a diameter equal to the picture height and within the 2 percent circles (±1.0 percent) in the remainder of the raster.

Before the advent of color, competitive cost pressures caused manufacturers to design receivers with marginal scanning linearity and geometric distortion. Color, however, imposed a requirement for the registration of the three colors that could not be compromised. Eliminating registration errors also led to improved linearity and reduced

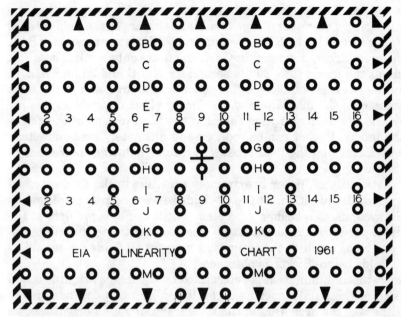

Figure 2.19 EIA linearity test chart.

geometric distortion. Typical geometric distortion specifications for production receivers are as follows:[5]

Aspect ratio	± 5 percent
Overscan	5–10 percent
Vertical and horizontal linearity	5–8 percent
Parallelogram at each corner	90° ± 1°

The vertical and horizontal linearity of 5 to 8 percent is to be compared with the typical 1 percent tolerance for cameras (see above).

2.17.5 Flicker

Flicker is an inherent defect that results from the sampling technique that is used in the production and display of television signals. It can be minimized or eliminated by the use of a sufficiently high repetition rate. Both television and motion pictures depend on the retentivity of the eye to merge a rapid sequence of images into a single continuous one. If the repetition rate of the images is too low, the eye will fail to merge them, and flicker results. The lowest repetition rate at which a continuous image is perceived is called the *critical fusion frequency*.

Two types of flicker are encountered in television systems that employ interlaced scanning. *Large-area flicker* occurs at the field rate, 60 fields per second in the NTSC system and 50 fields per second for PAL. *Small-area flicker,* described in Chap. 1, affects only vertical detail. It occurs at the frame rate, 30 per second for NTSC and 25 for PAL.

Large-area flicker involves the entire image area and is the most troublesome. The magnitude of the flicker effect depends on two parameters, the field repetition rate and the illumination of the retina by the image. Flicker can be reduced and ultimately made unnoticeable by increasing the field repetition rate or reducing the retinal illumination.

The visibility of flicker is extremely sensitive to the field rate and the retinal illumination. At the 60-fields-per-second rate employed by NTSC systems, it is seldom a problem, but it can be a problem at the 50-fields-per-second rate. One study showed that the retinal illumination can be as much as three times as great at 60 fields per second as at 50 fields before flicker is observed.[6]

The retinal illumination is determined by the image brightness and the size of the pupil's aperture. The latter, in turn, depends on the brightness of the area around the image, the *surround brightness,* as well as that of the image itself. In a darkened motion picture theater, the surround brightness is low, the pupil aperture is relatively large as the eye automatically adjusts to the low light level, and the retinal illumination is high in relation to the screen brightness. In a typical home environment, the surround brightness is higher, the pupil aperture is smaller, and the retinal illumination is lower in relation to the image brightness. As a result, the image brightness can be greater without causing visible flicker in a home TV set than with motion pictures in a theater. The difference is increased still further by the use of the lower field rate of 48 per second in motion picture systems.

The image brightnesses that can be tolerated without flicker under typical viewing conditions, relative to a 60-field-per-second NTSC system, have been estimated to be as shown in Table 2.7.[6]

Large-area flicker has not been a problem with 60-field-per-second NTSC systems with the image and surround brightnesses that are normally encountered. Looking ahead to future HDTV systems, it is a

TABLE 2.7 Flicker-Brightness Tolerance

	Field Rate, fields/s	Relative Brightness, %
NTSC TV	60	100
PAL TV	50	29
Theater movies	48	20
Home movies	48	20

potential problem. In an effort to increase the contrast ratio, viewing conditions with a lower surround brightness may be chosen and the white field brightness increased. Both of these conditions would increase the vulnerability to flicker. Also, the use of a wide screen and lower viewing ratio places the edges of the screen on the periphery of vision, where flicker is more noticeable (presumably one of nature's natural protective functions to make the individual aware of threatening movements out of the line of vision).

Small-area flicker affects only vertical detail, i.e., horizontal edges, and it occurs at a rate of 30 or 25 Hz in systems employing interlaced scanning. It has not been a problem with broadcast systems because the vertical definition is not great enough to produce sufficiently sharp edges. The high numbers of scanning lines proposed for HDTV systems significantly reduce this problem, and the use of sequential scanning would eliminate it. It is doubtful, however, that small-area flicker is serious enough to warrant the increase in bandwidth.

2.17.6 Aliasing

Aliasing is a defect that results from sampling, the transmission and display of scenes that are continuous in space or time by means of series of discrete samples. Brightness variations in the vertical dimension of a scene are sampled by the scanning lines, and motion in the scene is sampled by a series of still frames. An additional sampling process, rapid time samples of the signal amplitude, is also employed in digital systems (see Chap. 4).

In spite of the aliasing problem, sampling is a fundamental technology that is necessary for the production, recording, transmission, and display of television images. Scanning line and frame aliasing are described in this chapter, together with the theory of sampling.

2.17.6.1 Scanning line aliasing. Vertical brightness variations are sampled by the horizontal scanning lines, and if the dimensions of the vertical detail are comparable with the scanning line spacing, aliasing will occur.

The result is a third and spurious set of lines having a line number equal to the difference between the number of scanning lines in the raster and the line number of the lines in the image. With typical subject matter, this produces an objectionable moiré pattern in the image. (See Fig. 2.20 for a similar effect with time domain sampling.) With typical subject matter, this produces an objectionable moiré pattern in the image.

This form of aliasing is particularly visible with subject matter that has a strong geometric pattern, e.g., a herringbone suit. Motion in the

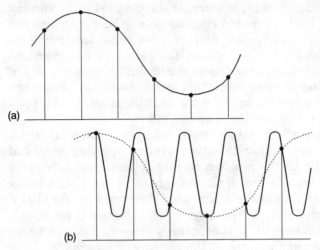

Figure 2.20 Sampling—time domain.

picture also enhances the effect, and a geometric pattern may appear to twinkle. It is also very noticeable in sharp, high-definition images. It was not a serious problem in early television systems because the fine detail did not have sufficient amplitude to interact with the scanning line pattern. With the improved definition of current broadcast systems, it has become a minor problem, and it is a major consideration in HDTV system design.

2.17.6.2 Frame aliasing. Frame aliasing occurs when one cycle of a repetitive horizontal pattern moves vertically more than one scanning line during the frame interval. For example, if a fine-grained repetitive pattern moves *upward* 1.5 lines during the frame interval, the visual effect will be the same as if it had moved *downward* 0.5 line. This is the cause of the motion picture phenomenon in which wagon wheel spokes appear to be turning backward.

Frame aliasing is an inherent sampling defect, and can be reduced or eliminated only by increasing the frame repetition rate.

2.18 Analysis of the Sampling Process

The usage of sampling extends beyond the scanning line and frame sampling described in Sec. 2.17.6, and its theory has been the subject of intensive analysis. Although the functions that are subject to sampling are not always electrical, the analysis is often carried out by the use of electrical analogs. As an example, frequency becomes the analog of television line number. The powerful mathematical tools that have been developed for electric circuit analysis are then available. The analysis has been carried out in both the time and frequency domains, because both are needed for a full description of the process. In addition, the analysis covers sampling in the image dimensions, *spatial sampling,* and in time, *temporal sampling.*

Applications of sampling in analog television include the following:

Parameter	Sampling Process
Vertical detail	Scanning line
Image motion	Frame
Horizontal detail	CCD imagers
	Shadow mask kinescopes
	Digital signals

Sampling is an essential ingredient of digital television systems (see Chap. 4), and in these systems the image is sampled horizontally as well as vertically. In a related process, the amplitude of the signal is *quantized,* i.e., it is represented by successive discrete levels rather than by a continuous function. Thus digital systems sample the image in both dimensions and quantize the signal amplitude. Sampling and signal amplitude quantizing as employed in digital systems are described in Chap. 4.

2.18.1 Sampling analysis—time domain

The sampling process in the time domain is shown in Fig. 2.20. In Fig. 2.20*a*, the sampling rate is more than twice the signal frequency, and in Fig. 2.20*b*, it is less. A sampling rate of twice the highest signal frequency component is known as the *Nyquist frequency,* or the *Nyquist limit*; lower sampling rates are known as *sub-Nyquist sampling.* Sub-Nyquist sampling generates a third frequency component, equal in frequency to the difference between the sampling and signal frequencies (see Fig. 2.20) and analogous to a beat frequency between two audio tones. It is the beat frequency that produces aliasing and demonstrates the Nyquist criterion—the sampling frequency or rate should be at least twice the highest signal frequency component.

For vertical detail, the sampling rate is the number of scanning lines in the raster, and ideally this would equal the highest line number of the vertical signal components generated by the electrooptical components (see Fig. 2.12). (Because the line number measurement counts both black and white lines, there are two television lines per cycle.) This would eliminate scanning line aliasing. The aperture response of earlier cameras at line numbers above the Nyquist limit was so low that aliasing was seldom a problem. But in modern broadcast systems using cameras with extended aperture response, the amplitude of the signal components above the Nyquist limit can be significant. It is for this reason that aliasing becomes a problem with scenes that have a strong geometric pattern.

2.18.2 Sampling analysis— frequency domain

Sampling pulses have the characteristics of an RF carrier that is amplitude-modulated by the signal. The frequency spectrum of the pulses consists of a strong carrier component at the sampling frequency with sidebands that correspond to the frequency components in the sampled signal. (The sampling pulses themselves are rich in harmonics, and the pattern of carrier and sidebands is repeated at harmonics of the sampling frequency.) The resulting frequency spectra for baseband and sampled signal are shown in Fig. 2.21. Figure 2.21a shows the spectra when the sampling rate is more than twice the highest baseband frequency and thus meets the Nyquist criterion. Figure 2.21b shows the effect of a sub-Nyquist sampling rate. In this case, there is an overlap between the frequency spectra of the baseband and the sampled signal. This overlap cannot be removed by subsequent signal processing, and it is the source of aliasing.

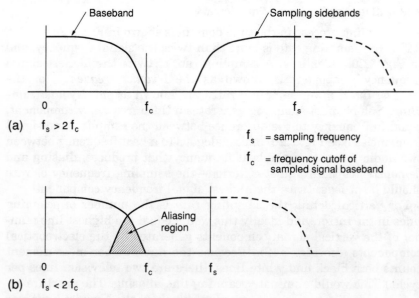

Figure 2.21 Sampling—frequency domain.

An analysis of the frequency spectra shows that their overlap and the resulting aliasing do not occur if the sampling rate meets the Nyquist criterion.

References

1. Otto H. Schade, Jr., *Image Quality, A Comparison of Photographic and Television Systems*, Princeton, N.J., RCA Laboratories, 1975.
2. Otto H. Schade, Jr., "Electro-Optical Characteristics of Television Systems," *The RCA Review*, Vol. 9, March, June, September, December 1948.
3. S. Herman, "The Design of Color Television Color Rendition," *J. SMPTE*, Vol. 84, 1975.
4. "Engineering Aspects of Television Allocations," TASO Report to the FCC, March 1, 1977.
5. Blair K. Benson, *Television Engineering Handbook*, Chap. 13, McGraw-Hill, New York, 1992.
6. Blair K. Benson, loc. cit.

Color Video Signals

3.1 Introduction

This chapter expands the description of color video signals contained in Chap. 1. The first half is devoted to colorimetry* and the properties of color that are the basis of color television technology. The second half describes the signal formats that have been developed to transmit and record color images.

3.2 The Properties of Color—A Summary

The term *color* as used in color television, has a dual meaning—it is both a physical property of visible light and the perception of this property by human vision. Color as a physical property can be defined and measured in objective and precise terms. Its perception is subjective: it varies with individuals, and it depends on the surrounding environment, so that it must be specified less precisely on the basis of an average or "standard" observer under "standard" conditions.

As with other aspects of image quality, a major task of television engineers is to establish and define measurable criteria for color images that coincide as closely as possible with people's perception.

Each picture element in a color image has three basic properties. In objective terms, they are luminance, hue, and saturation. The corresponding perceptual terms are brightness, color, and purity.

The term *luminance* has approximately the same meaning in color television as in monochrome (see Secs. 1.8 and 2.2.2). Like a monochrome system, a color television system must have the capability of

---

*Colorimetry originally meant the measurement of color with a colorimeter, but it now has the broader meaning of the science of color.

transmitting the luminance of each picture element, but in addition, it must transmit and reproduce each element's hue and color purity or saturation.

3.2.1 Hue

The most prominent characteristic of color picture elements is their hue, the sensation created by visible light that is commonly described as its color. The term *hue* is often used synonymously with "color."

Visible light is electromagnetic radiation having a spectrum of wavelengths* extending from approximately 400 to 780 nm (1 nanometer = 10^{-9} meter),† and its hue is determined by its wavelength. The band centers of the wavelengths that produce the sensation of common hues are presented in Table 3.1.

TABLE 3.1   Hue vs. Wavelength

Hue	Approximate wavelength, nm
Violet	400
Blue	450
Cyan	490
Green	520
Yellow	575
Orange	590
Red	640

This table is based on monochromatic light, i.e., light of a single wavelength. With the important exceptions of pure spectral colors resulting from the excitation of atoms, as in a sodium vapor light or a laser, the colors that appear in nature are polychromatic, i.e., they are a mixture of many wavelengths. Each has a dominant wavelength that establishes the visual perception of its hue, but it may contain radiation components with wavelengths that extend over the entire visible spectrum.

White or gray light results when the radiation at all wavelengths is present in approximately equal amounts.

Radiation with wavelengths just beyond the limits of the visible spectrum also has important applications in television. In the range

---

*Wavelength rather than frequency is used in the visible and neighboring regions of the spectrum. Wavelength and frequency are related by the equation.

[Wavelength (m)][frequency (Hz)] = 3×10^8 m/s (the velocity of light)

†Scientists sometimes use the Angstrom unit, 10^{-10} m, rather than the nanometer for specifying wavelength.

from approximately 780 to 25,000 nm it is known as near infrared, while in the range of 100 to 400 nm it is known as ultraviolet.

3.2.2 Saturation

The appearance of any color can be duplicated by a mixture of white or gray light and a pure spectral color, the dominant wavelength, in the proper proportions (see Sec. 3.5). The ratio of the magnitude of the energy in the spectral component to the total energy of the light defines its purity or *saturation*. A pure spectral color has a saturation of 100 percent, while the saturation of white or gray light is zero.

3.3 Color Specifications

The three basic physical properties of color radiation—luminance, hue, and saturation—can be completely specified by its spectral distribution, a plot of its radiant energy vs. wavelength. Figure 3.1 shows the spectral distributions of incident light on a reflecting surface, the reflectivity of the surface, and the reflected light, which is the product of the first two.

A graph of amplitude vs. wavelength, however, is not always the most useful format for specifying color—numerical criteria are often more convenient—and it does not describe the relationship between the physical properties of color and their visual perception.

Scientists have studied the nature of this relationship for the past

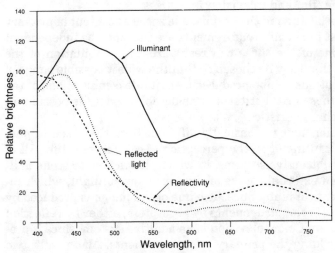

Figure 3.1 Spectral distribution of reflected light.

400 years (Sir Isaac Newton carried out basic research on this subject in the early seventeenth century), and they have developed numerical criteria for specifying color that are stated in technical terms but are based on the subjective perceptions of thousands of viewers. Four of these criteria are commonly used in color television: (1) mix of primary colors, (2) dominant wavelength and saturation, (3) luminance and color differences, and (4) coordinates on the CIE Chromaticity Diagram.

All of these criteria directly or indirectly specify three independent variables. This is consistent with the trichromatic theory of color, which states that the sensation of color results from the stimulation of three sets of cones in the retina, each with a different spectral sensitivity. The perceived hue and saturation of a color are determined by the ratios of the amplitudes of the responses of the sets of cones to the color stimulation. Thus two colors will appear to have the same hue and saturation if they stimulate the same response from the retinal cones, even if the distribution of their spectral energies is different. (Their *appearance*, however, is also affected by differences in the viewing environment, a fact well known to painters.)

3.4 The Primary Colors

A consequence of the trichromatic theory of color is that the hue and saturation of most colors can be duplicated by combining three primary colors in the proper ratio. This theory was one of the results of Sir Isaac Newton's work. It has since been developed further to meet the needs of the film and television industries.

There is wide latitude in the permissible spectral content of primary colors. The most important requirements are that they be independent (i.e., that no one of the three can result from a combination of the other two) and that they be chosen so that the widest possible range of hues and saturations can be produced by suitable combinations of the three. Within these broad criteria, considerable variety is possible in their spectral characteristics.

Television colorimetry is basically different from that used in photography and painting or in the perception of color in real life. All of the latter are subtractive systems, in which the picture or scene is illuminated by an external source of light, such as sunlight, which includes components of many hues. The hue of the image is produced by the subtraction of color components by absorption, either by reflection or by transmission. Color television is an additive system, since it produces hues by adding the primary color components. As a result, two types of primaries, subtractive and additive, have been defined.

3.4.1 Subtractive primaries

The subtractive primaries are magenta, yellow, and cyan. Each absorbs one of the additive primaries and reflects or transmits the other two (see Table 3.2).

TABLE 3.2   Subtractive Primaries

Primary	Reflects or transmits	Absorbs
Magenta	Red and blue	Green
Yellow	Red and green	Blue
Cyan	Blue and green	Red

Since magenta has a reddish cast and cyan is bluish, it is popularly (and erroneously) stated that the primary colors are red, yellow, and blue.

3.4.2 Additive primaries

The additive primaries (Table 3.3) can be derived from the subtractive; for example, red can be produced by mixing magenta and yellow, the former absorbing green and the latter blue from the incident white illuminant.

TABLE 3.3   Additive Primaries

Combinations	Hue of combination
Red plus green	Yellow
Blue plus green	Cyan
Blue plus red	Magenta

The luminance of the additive primary colors, R (red), G (green), and B (blue), and the amplitude of their electrical analogs, E_R, E_G, and E_B, are the basic color television parameters, and all other color specifications are derived from them. [After gamma correction (see Sec. 2.15.5), it is customary to designate the amplitude of the electrical signals as E_R', E_G', and E_B'.] The outputs of the camera and the inputs to the kinescope are the electrical analogs of the three primary colors in the scene.

Additive primaries were seldom used until the advent of color television because it is virtually the only imaging system that creates pictures by the addition of color components. The additive primaries are red, green, and blue. Other hues, including the subtractive primaries, can be produced by combining them.

It is not intuitively obvious that yellow light is produced by mixing red and green(!), but it can be confirmed by superimposing red and green spotlights on a screen.

3.5 Dominant Wavelength and Saturation

It can be shown, both by theory and by experiment, that the luminance, hue, and saturation of every color in nature, no matter how complex its spectral distribution, can be duplicated by a mixture of white or gray light and monochromatic light in the proper proportions. The wavelength of the monochromatic component of this mixture that establishes its perceived hue is known as the dominant wavelength.

The electrical analogs of the dominant wavelength and saturation are the basis for the NTSC, PAL, and SECAM systems that are used for the transmission of color by all present-day broadcast systems. See Sec. 3.10.1.

3.6 Luminance and Color Difference Components

Colors can be specified in terms of their *luminance, Y,* and *color difference, R – Y* and *B – Y,* components. The amplitudes of these components are related to the primary amplitudes by matrix equations (see Sec. 3.12). Their electrical analogs are the basis for *component* recording and transmission, a format that is widely used for video recording and other applications with high demands for quality that do not require compatibility with broadcast transmission standards.

3.7 The CIE Chromaticity Coordinates

The CIE (Comité Internationale de l'Eclairage) chromaticity coordinates are based on a system for describing colors in terms of the color matching functions x', y', and z', shown in Fig. 3.2. These functions were developed on the basis of hundreds of observations, and are used to calculate the *tristimulus values X, Y,* and *Z:*

$$X = 380 \int_{380}^{780} L(\lambda)x'(\lambda)d\lambda \qquad (3.1)$$

$$Y = 380 \int_{380}^{780} L(\lambda)y'(\lambda)d\lambda \qquad (3.2)$$

$$Z = 380 \int_{380}^{780} L(\lambda)z'(\lambda)d\lambda \qquad (3.3)$$

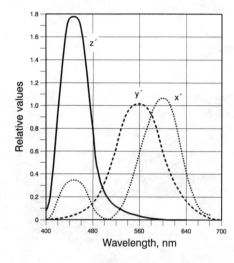

Figure 3.2 Color matching functions.

where $L(\lambda)$ is the spectral energy density and x', y', and z' are the color matching functions.

The properties of the color matching functions are such that colors with the same tristimulus values will appear the same, even though their spectral distributions are different.

The chromaticity coordinates, x and y, are calculated from the tristimulus values:

$$x = \frac{X}{X + Y + Z} \qquad (3.4)$$

and

$$y = \frac{Y}{X + Y + Z} \qquad (3.5)$$

(Since $x + y + z = 1$ by definition, it is unnecessary to calculate z.)

The concept of chromaticity coordinates was a major milestone in the development of imaging system technology. It was developed during the late 1920s and early 1930s, and the coordinates of the colors in the visible spectrum were first calculated and published in 1931. It is an invaluable tool for the analysis and design of color television products and systems. It is also a link between the objective physical nature of color and its subjective perception.

The plot of the chromaticity coordinates of visible light is known as the CIE Chromaticity Diagram (see Fig. 3.3). The color matching functions and the Chromaticity Diagram are known collectively as the *CIE 1931 Standard Observer*. A supplement to the 1931 paper, the *CIE 1964 Supplementary Standard Observer*, added data for viewing angles greater than 2°.

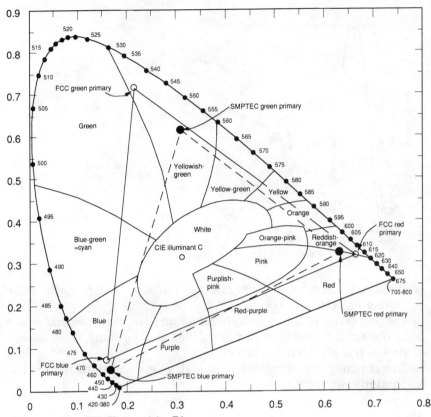

Figure 3.3   The CIE Chromaticity Diagram.

The following features of the Chromaticity Diagram should be especially noted:

1. The coordinates of spectral (monochromatic) colors are located on the horseshoe-shaped curve around the periphery of the diagram. Their wavelengths are indicated on the curve.

2. The coordinates of white light are located in the center of the diagram. The area on the graph corresponding to white light is rather large because there is an almost infinite variety of white hues. The three sets of coordinates for white light that have been established by the CIE are tabulated in Table 3.4. The spectral power distributions of these illuminants are shown in Fig. 3.4.

3. All visible colors are located within the area bounded by the spectral curve and the line (the alychne) joining its red and blue ends.

TABLE 3.4   Chromaticity Coordinates of Standard Illuminants

Designation	Source	x	y
Illuminant A	Tungsten at 2856 K	0.4476	0.4074
Illuminant C	Daylight	0.3101	0.3516
Illuminant D_{65}	Daylight (revised)	0.3127	0.3290

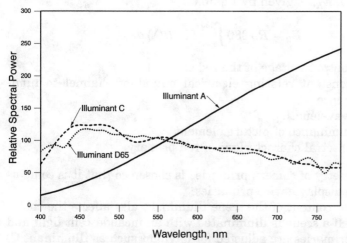

Figure 3.4   Spectral distribution of standard illuminants.

4. Every color whose coordinates lie on a line joining a point at the white center of the diagram to the outer spectral curve have the same hue. The dominant wavelength for the color is determined by the point at which the line intersects the spectral curve. Its saturation is indicated on its location along this line, ranging from zero at the center to 100 percent at the spectral curve.

5. The coordinates of primary sets are located at the corners of a triangle. All colors that can be reproduced by a given set of primaries have coordinates within the triangle.

6. The specification of color primaries is one of the most important applications of chromaticity coordinates. See Sec. 3.8.1.

3.8   Camera Colorimetry

3.8.1   Camera primaries

The camera primary signals are the electrical analogs of the red, green, and blue visual components of the scene. Their colorimetry is specified by a graph of response vs. wavelength. It is determined initially by the filters in the optical color splitter, but it can also be modified electrically in a camera matrix circuit by a process known as *masking*; see Sec. 3.8.2. With masking it is possible to achieve a spectral response that would be impossible with optical filters alone.

The amplitude of each primary signal is determined by the product of the spectral content of the scene and the colorimetry of the primary integrated over the entire spectrum. For example, the amplitude of the red signal, E_R, is given by Eq. (3.6):

$$E_R = R_C \, 380 \int_{380}^{780} L(\lambda)R(\lambda) \, d\lambda \qquad (3.6)$$

where E_R = output voltage of the red channel
  R_C = constant relating electrical output of channel to luminance
  λ = wavelength
  $L(\lambda)$ = luminance of picture element
  $R\lambda)$ = spectral characteristic of red primary

The colorimetry of camera primaries is chosen so that it is compatible with the display device primaries.

The spectral content of the scene illuminant also affects the choice of primaries. If a scene is illuminated with an incandescent light and the camera primaries are adjusted for daylight such as illuminant C, the reproduced image will appear reddish. Compensation for effect is achieved by readjusting the camera matrix circuits, and a switch can be provided for the matrix circuits with two positions, one for illuminant C and one for incandescent light.

The camera primaries recommended by the NTSC and adopted by the FCC in 1953 were based on compatibility with the NTSC recommendations for display device primaries and illuminant C. Their spectral responses are shown in Fig. 3.5.

These primaries contain negative lobes that cannot be produced optically. This problem has two solutions. One is to use primaries that ignore the negative lobes and the sidelobe in the red primary, as shown in the cross-hatched areas in Fig. 3.5. This is a compromise that produces satisfactory results, but with some color distortion. The other is to employ matrix circuits to create the negative lobes electrically; see Fig. 3.6.

Another departure from the NTSC/FCC standards is required by the industry practice, described in Sec. 3.9, of employing receiver primaries that deviate considerably from the original FCC specifications in order to provide greater screen brightness. This requires a corresponding modification in the camera primaries for best results.

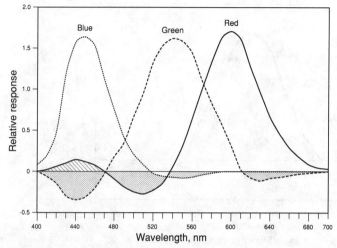

Figure 3.5 NTSC/FCC standard camera primaries.

3.8.2 Matrix circuits

Electric matrix circuits convert an array of coefficients of independent variables into another array through a mathematical transform. They can be used to alter the spectral response of camera primaries, either to provide negative lobes that cannot be produced optically or to compensate for changes in the scene illuminant.

The operation of these circuits is illustrated by the example of the NTSC green primary shown in Fig. 3.6. Optical filters cannot produce the negative values, but they can be produced electrically by a matrixing operation:

$$E_{GM} = E_G - 0.2E_B - 0.05E_R \tag{3.7}$$

Figure 3.6 shows the close approximation of the matrixed signal to the NTSC primary.

This is a simple example involving only the addition and subtraction of linear functions. More complex circuits make use of nonlinear elements so that the matrix coefficients are dependent on the signal levels. With these systems, the operator can "paint" the picture in accordance with esthetic judgments. Also, they make the precise matching of cameras possible.

3.9 Display Device Primaries

It is desirable to standardize the primaries of color kinescopes to provide a uniform basis for the design of matching camera primaries. The

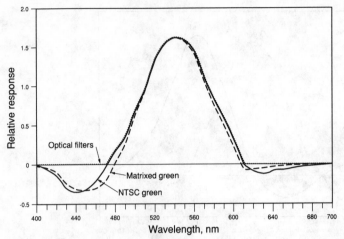

Figure 3.6   Green primaries—effect of matrixing.

colorimetry of kinescope primaries is determined by the characteristics of the phosphors used in the picture tube and is specified by the phosphors' CIE coordinates.

3.9.1   FCC/NTSC primaries

Chromaticity coordinates of display device primaries were proposed by the NTSC and adopted by the FCC in 1953. Their values are given in Table 3.5 and plotted in Fig. 3.3.

TABLE 3.5   FCC/NTSC Chromaticity Coordinates (1953)

	x	y
Red	0.67	0.33
Green	0.21	0.71
Blue	0.14	0.08

3.9.2   SMPTE Standard C primaries

With growing experience in receiver manufacture, it was found that greater screen brightness could be obtained with primaries that differed from the original NTSC recommendations. Customers preferred the brighter pictures, and many receivers were manufactured with the revised primaries. In recognition of this reality, the SMPTE (Society of Motion Picture and Television Engineers) issued revised primary chromaticity coordinates in 1982. They are given in Table 3.6.

TABLE 3.6   SMPTE Standard C Primaries (1982)

	x	y
Red	0.635	0.340
Green	0.305	0.595
Blue	0.155	0.070

The SMPTE coordinates are also plotted in Fig. 3.3, which shows that the added brightness is obtained at the expense of a more limited gamut of green, yellow, and red hues but a slightly greater gamut of blues. The compromise is not serious, however, as illustrated in Fig. 3.7,[1] which compares the gamut of spectral responses of typical paints and dyes with Standard C. The color gamut of the television system, the area within the triangle, compares favorably with the gamut of these spectral responses.

3.10   Video Signal Formats for Color

All video signal formats for color are derived from the primary signals E_R, E_G, and E_B or their gamma-corrected alternatives E_R', E_G', and E_B', which are generated by the camera chain. The camera signals are in analog form and have amplitudes proportional to the spectral energy for the primary colors in the scene. At the receiver, the inputs to the receiver kinescope or other display device are modified replicas of these signals. Between the camera chain and the kinescope input, however, the signal is usually transformed to other formats, either analog or digital, that are better suited to the functions of signal processing, recording, transmission, and broadcasting. This chapter is devoted to a description of analog formats; digital formats are described in Chaps. 4 and 6.

For all formats, a luminance signal component E_Y' is generated by matrix circuits in accordance with Eq. (3.8):

$$E_Y' = 0.299E_R' + 0.587E_G' + 0.114E_B'   \tag{3.8}$$

The coefficients of the components E_R', E_G', and E_B' are based on the relative sensitivity of the human eye to the primary colors. The magnitude of E_Y' is then proportional to the perceived brightness of the scene.

Video signal transmission and recording formats can be divided into two classes, *composite* and *component*. In composite formats, the luminance and chrominance signals are multiplexed within a single channel or on the same carrier.

Composite color formats can be either *compatible* or *noncompatible*

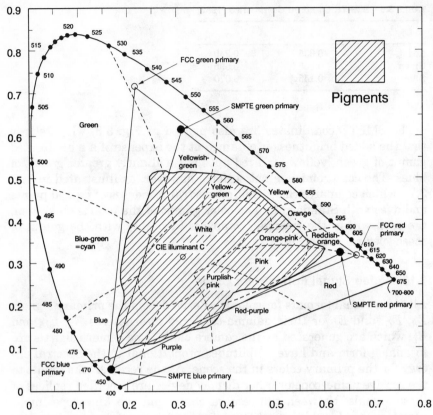

Figure 3.7   Color gamuts of dyes and paints. (*From K. Blair Benson*, Television Engineering Handbook, *McGraw-Hill, New York, 1986, Fig. 22-18.*)

with the monochrome format. Compatible color transmissions can be received on monochrome receivers, and monochrome transmissions can be received on color receivers. All formats currently used for broadcasting are compatible.

In component formats, the luminance and chrominance signals are transmitted or recorded on separate channels or at separate times, as in the MAC systems. They are widely used for recording in complex program production systems where the highest quality must be achieved, even though the final copy may require several generations of recordings. They are also used for satellite transmission.

3.10.1   Composite formats

Compatible composite formats used for broadcasting and related services must meet four requirements:

1. The luminance and chrominance information must be contained in a single channel (hence they are composite).

2. The luminance component must be sufficiently similar to the monochrome signal that color broadcasts can be received on monochrome receivers and vice versa. This is the *compatibility* requirement.

3. The bandwidth must be no greater than that required for monochrome.

4. The format must have the approval of the appropriate regulatory authority (the FCC in the United States).

Three compatible composite formats are in wide use for broadcasting: NTSC (for National Television Systems Committee) in the United States, Canada, Mexico, and Japan; PAL (for Phase Alternating Lines) in western Europe (except France); and SECAM (Sequentiel Couleur avec Memoire) in France, eastern Europe, and the former USSR. All meet these requirements. They are similar in that they employ a subcarrier or subcarriers to specify the saturation and dominant wavelengths of the picture elements. The NTSC and PAL systems employ a single phase- and amplitude-modulated subcarrier that is superimposed on the luminance signal (see Secs. 3.12.2 and 3.12.3). The SECAM system employs two frequency-modulated subcarriers.

3.10.2 Component formats

A major disadvantage of subcarrier composite systems is the impossibility of completely eliminating crosstalk between the luminance and chrominance components as the result of nonlinearities in the system and overlaps of the sideband spectra of the subcarriers. The nonlinearities cause *differential gain* and *differential phase* distortion of the chroma component and visibility of the subcarrier in the luminance component. The spectrum overlaps cause *cross-color* and *cross-luminance intermodulation*. Differential gain and phase distortions are significant problems with broadcast TV systems (see Secs. 3.17.3 and 3.17.4). Cross-color results when the luminance signal contains frequency components near the subcarrier frequency, and spurious colors are generated in the image. Cross-luminance occurs when the signal representing a given point changes from field to field or on the edges of objects so that the subcarrier cancellation is not complete. A dot pattern then occurs in the luminance channel.

Another disadvantage of subcarrier composite systems results from the fact that the noise in transmission systems that use frequency modulation, e.g., microwave and satellites, increases linearly with the modulating frequency. This puts the subcarrier at the high-noise end of the spectrum and can lead to a poor signal-to-noise ratio for the chrominance signal.

Signals employed for editing, point-to-point transmission, and recording are subject to none of the constraints of broadcasting, and it is possible to choose a component format that eliminates the use of a subcarrier with its crosstalk and noise problems. Component formats have no cross-luminance or cross-chroma distortion. The most common component formats employ three components—luminance, or E_Y, and the color difference components, $E_R - E_Y$ and $E_B - E_Y$. They may be transmitted on separate channels or time-multiplexed on a single channel as with MAC.

3.11 Analog Color Television System Configurations

Figure 3.8 is a simplified diagram that shows the configurations of composite and component color television systems. Both begin with a camera chain that includes a transducer for each primary color (a tube or solid-state device) and signal processing circuitry that generates the primary signals, E_R', E_G', and E_B'. The processing functions include white and black level control, aperture correction and image enhancement, color correction by matrixing, gamma correction, and sync and blanking insertion.

In composite systems, the camera chain is followed by an encoder that produces the composite signal by multiplexing the primary signals.

This composite signal is transmitted to the receiver, which decodes the signal to recover the original E_R', E_G', and E_B' primary signals; these are the inputs to the picture tube.

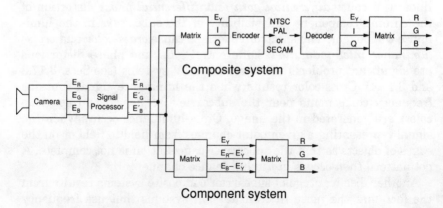

Figure 3.8 Color television system functional diagram.

In a component system, the luminance and color difference signals are generated by matrixing the primary signals. After matrixing, the component signals may go to a video recorder; to a microwave, fiber optic, or satellite transmission system; to a MAC time-multiplex circuit; or to an A-to-D converter for conversion to a digital format.

3.12 Broadcast Signal Components

3.12.1 The luminance signal component

The amplitude of the luminance signal component, E'_Y, is given by Eq. (3.8). Its bandwidth varies from country to country; see Table 3.7.

TABLE 3.7 Color Television System Bandwidths

Country	System	Bandwidth, MHz
United States, Japan	NTSC	4.2
Canada, Mexico	NTSC	4.2
Great Britain	PAL	5.5
Germany, Austria, Italy	PAL	5.0
France	SECAM	6.0
Former USSR	SECAM	6.0

Most European countries use UHF for color broadcasting, which accounts for the greater bandwidths available for their systems.

3.12.2 NTSC chrominance signal components

It is possible to conserve spectrum space by deriving two chrominance components, E'_I and E'_Q, called the I and Q signals, from the primary components:

$$E'_I = 0.60E'_R - 0.28E'_G - 0.32E'_B \tag{3.9}$$

$$E'_Q = 0.21E'_R - 0.51E'_G + 0.30E'_B \tag{3.10}$$

The chrominance components, E'_I and E'_Q, have the following characteristics:

1. E'_I is at a maximum for orange or cyan hues, while E'_Q is at a maximum for green or magenta.

2. The ability of the eye to discern fine detail is less in colors than for monochrome, and it is less for magenta than for orange. As noted

TABLE 3.8   Component Bandwidths

Component	Bandwidth, MHz
E_Y' (luminance)	4.2
E_I' (orange-cyan)	1.5
E_Q' (green-magenta)	0.5

above, this property is utilized in the NTSC to reduce the bandwidths of the chrominance signal components; see Table 3.8.

3. With white light, $E_R' = E_G' = E_B'$, and all chrominance components, E_I' and E_Q' or $E_R' - E_Y'$ and $E_B' - E_Y'$, disappear.

4. Their amplitudes relative to the luminance signal are a tradeoff between a large value that would lead to excessive crosstalk with the luminance component and a small value that would lead to an inadequate signal-to-chroma noise ratio.

3.12.3   PAL chrominance signal components

With video bandwidths of 5 to 5.5 MHz available, PAL systems do not need to be as frugal in their use of spectrum space as the NTSC system. Accordingly, there is no need to add to the complexity of PAL systems by creating the I and Q signals, and the color difference primaries are used directly for producing the color subcarrier.

The two chrominance components in the PAL system are E_U' and E_V'. Their equations are

$$E_U' = 0.493(E_B' - E_Y') \qquad (3.11)$$

$$E_V' = 0.877(E_R' - E_Y') \qquad (3.12)$$

The bandwidth of the E_U' and E_V' components is 1.3 MHz (to the -3-dB points).

3.13   Subcarriers

The essence of the differences between the NTSC and PAL color television systems is their choice of subcarrier frequencies, phases, and formats. The subcarrier frequencies in both systems are chosen to be as high as possible while keeping their sidebands within the video passband. Their relationship to the frequencies and phases of the scanning lines and fields must also minimize the crosstalk between the luminance signal and the subcarrier and the visibility of the subcarrier in the composite signal. The solutions to these problems differ for each system.

3.14   The NTSC System Subcarrier

3.14.1   The NTSC subcarrier format

The NTSC subcarrier format was the first to be developed (in the United States in the early 1950s), and it is the simplest. It combines amplitude and phase modulation, its amplitude determining the saturation of the image and its phase the hue. The use of phase to transmit hue requires the transmission of a reference phase that is supplied by the burst, a short train of energy at the subcarrier frequency with a fixed phase on the back porch of the blanking pulse. The burst is described in Sec. 1.3.

The NTSC subcarrier e_{scn} can be derived either from the I and Q signals or from the color difference signals $E_R' - E_Y'$ and $E_B' - E_Y'$.

If e_{scn} is derived from I and Q, its equation is

$$e_{scn} = E_Q' \sin(\omega t + 33°) + E_I' \cos(\omega t + 33°) \tag{3.13}$$

where $w = 2\pi f_{sc}$ and f_{sc} is the subcarrier frequency.

If it is derived from the color difference signals, its equation is

$$e_{scn} = 0.877[0.562(E_B' - E_Y') \sin \omega t + (E_R' - E_Y') \cos \omega t] \tag{3.14}$$

Note that the subcarrier vanishes for white light. Also note that its average value is zero, so that in a linear system, the luminance signal is unaffected by the chroma. This is known as the constant luminance principle. Stated otherwise, there is no crosstalk from chrominance to luminance unless there are nonlinearities in the system.

The chrominance signal e_{scn} and its terms can be represented by vectors with the sine and cosine functions in quadrature. This is shown in Fig. 3.9, which shows the phase relationship of these functions to the burst and the significance of the 33° angle in the equation.

The NTSC subcarrier format is subject to crosstalk from luminance to chrominance, resulting from nonlinearities in amplifiers, recorders, and transmission systems that cause variations in gain or phase shift with the brightness level of the luminance signal. These effects are known as differential gain and differential phase. Differential gain causes distortions of the image saturation, and differential phase distortions of its hue.

Differential phase shift is a particularly serious problem because the eye is very sensitive to hue distortions. After the introduction of NTSC, the Europeans, who were considering other systems, jokingly said that NTSC meant "never the same color." With the passage of time, advances in technology have greatly reduced the differential gain and phase of video amplifiers, transmission systems, and record-

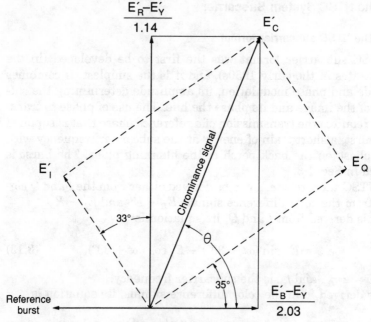

Figure 3.9 Vector diagram of the NTSC subcarrier.

ers, and their nonlinearities no longer need be significant problems. They still must be considered, however, in system design and operation, and the vulnerability of the NTSC format to system nonlinearities continues to be a potential problem.

Other forms of crosstalk are cross-color and cross-luminance; see Sec. 3.10.2.

3.14.2 NTSC subcarrier frequency

The NTSC subcarrier frequency is derived from the scanning line frequency, which in turn is an integral submultiple of the 4.5-MHz audio subcarrier used in broadcast systems. This relationship minimizes the beat between the audio and color subcarriers. It is also desirable to have the line frequency as close as possible to the monochrome line frequency of 15,750. As a result of these considerations, the line frequency is the 286th submultiple of 4.5 MHz:

$$f_{\text{line}} = \frac{4.5 \times 10^6}{286} = 15{,}734.26 \text{ Hz} \tag{3.15}$$

The field frequency is related to the line frequency by the number of scanning lines:

$$f_{\text{field}} = \frac{f_{\text{line}}}{525/2} = 59.94 \text{ Hz} \qquad (3.16)$$

It would be desirable for the field frequency to equal the frequency of the primary power source, 60 Hz (as for monochrome), but it is mathematically impossible to match this frequency precisely while other requirements are being met. The small difference between 59.94 Hz and 60 Hz causes hum bars to move slowly through the image, but this is not a serious problem with modern receivers because of their excellent filtering.

To reduce the visibility of the subcarrier in the luminance signal, its frequency is an odd multiple of one-half the line frequency. With this relationship, the dot pattern resulting from the subcarrier is stationary, and the phase of the subcarrier (for a given hue) at the beginning of each line reverses on successive scans. With the phase reversal, the dots on successive scans have opposite polarity and tend to cancel.

On the basis of these relationships, the equation for the subcarrier frequency f_{sc} is

$$f_{sc} = 455\left(\frac{f_{\text{line}}}{2}\right) = 3.579545 \text{ MHz} \qquad (3.17)$$

The relationship between subcarrier frequency and line rate must be maintained very precisely to avoid dot crawl. Usually the line rate is established by counting down from the subcarrier frequency by a series of frequency dividers:

$$f_{\text{line}} = \frac{2f_{sc}}{(5)(7)(13)} = 15,734 \, Hz \qquad (3.18)$$

Although the line and field rates are slightly different from those for monochrome television, the differences are so small that they cause no problems with monochrome receivers.

3.14.3 NTSC subcarrier reference phase

If 0° and 180° designate the subcarrier reference phases at the beginning of successive lines, their phases are:

Line	Field			
	1	2	3	4
1	0°		180°	
2		180°		0°
3	180°		0°	
4		0°		180°
5	0°		180°	

Note that four lines are required before the pattern repeats.

3.14.4 NTSC signal spectral content

Crosstalk of the chroma signal into the luminance channel, which produces a dot pattern in the image, is minimized by the choice of subcarrier frequency and phase. When the subcarrier and line frequencies are related, the dot patterns are stationary and thus less visible. And the reversal of the phase of the subcarrier on alternate frames causes at least a partial cancellation of the visual effect of the dots.

The video bandwidth requirements and the relative amplitude of the luminance and I and Q chrominance components over the video passband for a typical signal are shown in Fig. 3.10.

The luminance and chrominance components of the signal can occupy the same portion of the spectrum with a minimum of crosstalk because of their fine-grained spectral content. The content of the luminance component was shown in Fig. 1.8. It consists of groups of frequency components, each with a center frequency that is a multiple of the line rate, surrounded with a cluster of sidebandlike components separated by the field rate. Unless there is an unusual amount of motion in the picture, there are wide spectral gaps between these groups in which additional signal information can be inserted.

The spectrum of the subcarrier consists of similar groups of sideband components. Because of the integral relationship of the subcarrier frequency and the line rate in the NTSC system, the groups of subcarrier sidebands fit in the gaps in the luminance spectrum.

The interleaving of the luminance and chrominance components in the frequency domain is an indication that the chrominance signal will cause a minimum of crosstalk between the two signals. In the time domain this is evidenced by the fact that the subcarrier pattern is stationary and that it has opposite phases on successive scans.

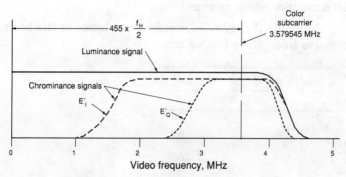

Figure 3.10 NTSC subcarrier frequency spectrum.

3.15 The PAL System Subcarrier

3.15.1 The PAL subcarrier format

The format of the PAL subcarrier was designed to minimize the effects of differential phase, one of the results of crosstalk from the luminance to the chrominance channel. It is similar to the NTSC format except that the phase of the subcarrier component derived from the color difference component $E'_R - E'_Y$ is shifted $\pm 90°$ at the end of each line with respect to the $E'_B - E'_Y$ component—hence the system's acronym, Phase Alternating Line rate.

Because of the wider frequency band used for most PAL systems, it is not necessary to employ the I and Q signals, with their greater complexity, and the subcarrier is derived directly from the color difference signals, $E'_R - E'_Y$ and $E'_B - E'_Y$. The equation for the PAL subcarrier is

$$e_{scp} = 0.493(E'_B - E'_Y) \sin \omega t \pm 0.877(E'_R - E'_Y) \cos \omega t \quad (3.19)$$

$$= E'_U \sin \omega t \pm E'_V \cos \omega t$$

The $\pm$ indicates the phase reversal on successive frames.

The phase relationship between the burst and the $E'_R - E'_Y$ and $E'_B - E'_Y$ subcarrier components is shown in vector form in Fig. 3.11. Unlike in NTSC, the PAL burst does not maintain a fixed relationship with the subcarrier components, but shifts 90° at the end of each line. This is sometimes called a swinging burst.

The dotted vectors show how the phase error due to differential phase is reversed on successive scans of the same line. For example, if differential phase causes a shift of the subcarrier phase and the cor-

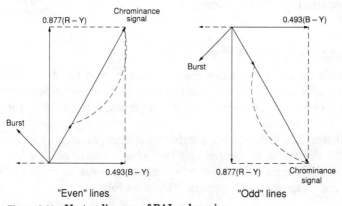

Figure 3.11 Vector diagram of PAL subcarrier.

responding image hue toward the $E'_R - E'_Y$ phase on the first scan, it will cause a shift toward the $E'_B - E'_Y$ phase in the next. Visual averaging of the two hues would then minimize the perceived hue distortion. This results in a number of annoying spurious effects. These can be avoided by electrical delay line averaging, in which the chrominance signal of each line is added to an electrically delayed signal of the same line from a previous scan; this technique, however, reduces the vertical chrominance resolution by a factor of 2.

As with NTSC, the subcarrier vanishes for white light.

3.15.2   PAL subcarrier frequency

In the PAL system, the field, frame, and line rates are the integral numbers used for European monochrome systems: the field and frame rates, f_{fr} and f_{fld}, are 25 and 50 Hz, and the scanning line rate, f_{line}, is 15,625 Hz, 625 times the frame rate.

Since two phase changes are involved in the PAL subcarrier—the phase of the reference carrier at the beginning of the line and the phase of the $E'_R - E'_Y$ component with respect to the $E'_B - E'_Y$—the relationship between line rate and subcarrier frequency used for NTSC would not be satisfactory for PAL. Instead, the subcarrier frequency is an odd multiple of one-quarter the line rate plus one-half the frame rate:

$$f_{scp} = 1135\left(\frac{f_{line}}{4}\right) + \frac{f_{fr}}{2} = 4.43361875 \text{ MHz} \qquad (3.20)$$

The term $f_{fr}/2$ causes the subcarrier phase on successive scans of the same line to shift by 90°.

A rigid relationship between the subcarrier frequency and the line rate is also required for PAL systems.

3.15.3   PAL subcarrier reference phase

The relationship between the subcarrier frequency and the line and frame rates results in a 90° phase shift in the subcarrier reference phase at the beginning of each line (as contrasted with 180° for NTSC).

If 0, 90, 180, and 270 indicate the carrier reference phase at the line beginning, and A and B represent the two phase relationships between the color difference subcarrier components, the phase relationships of the subcarrier and its components are as follows:

	Field							
Line no.	1	2	3	4	5	6	7	8
1	0A		90B		180A		270B	
2		270B		0A		90B		180A
3	90B		180A		270B		0A	
4		0A		90B		180A		270B
5	180A		270B		0A		90B	
6		90B		180A		270B		0A
7	270B		0A		90B		270B	
8		180A		270B		0A		90B

To achieve the proper phase of the burst for each line, it is necessary to shift its insertion time by one line at the field rate. This is done with a circuit known as a meander gate.

3.15.4 PAL signal spectral content

As a result of the relationship between the PAL subcarrier frequency and the line rate and the phase shift of the E_R' – E_Y' components at the end of each line, the PAL chrominance harmonics do not fall neatly between the luminance harmonics as with NTSC. In the time domain, the subcarrier phase shifts only 90° on alternate frames, and complete cancellation owing to phase reversal does not occur. Thus the subcarrier is inherently more visible in PAL than in NTSC, but the effect is mitigated by the higher PAL subcarrier frequency, which makes an uncanceled subcarrier less visible. The relationship between the luminance and chrominance spectrum components is shown in Fig. 3.12.

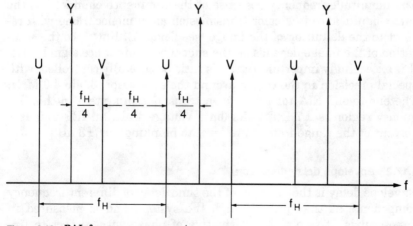

Figure 3.12 PAL frequency components.

3.16   Comparison of NTSC and PAL

PAL systems are usually used in countries where wider video bandwidth is available, often by the use of the UHF band for transmission. This provides a design flexibility that is not available to narrower-band NTSC systems. In addition, the wider bandwidth provides an advantage in signal quality that is independent of the choice of the color system.

PAL systems are sometimes described as being more *rugged,* meaning that they are less vulnerable to hue distortions resulting from phase errors in the transmission system.

On the other hand, NTSC systems are simpler, they do not require quite as much bandwidth, and their subcarriers are less visible. Also, improvements in transmission and recording systems make the greater vulnerability to phase errors less important.

3.17   Color Signal Distortions and Artifacts

Multiplexing the luminance and chrominance components of a color television signal with a subcarrier is not without problems, and certain distortions and artifacts (spurious signals produced artificially) result. This section describes some of the most significant that occur in NTSC and PAL systems. More detailed descriptions of these distortions, together with suggested tolerances and methods of measurement, have been published by the Electronic Industries Association, the FCC, and the CCIR.[2]

3.17.1   Chrominance-to-luminance gain inequality

The amplitude-frequency response of the luminance channel has the same significance in a color transmission as in monochrome with respect to the definition of the image (see Chap. 1). But since the saturation of the image depends on the subcarrier/luminance signal ratio, it is exceedingly important that this ratio be carefully controlled, with special precision in the region around the subcarrier, 3.0 to 4.0 MHz. The end-to-end EIA tolerance for a transmission system in this frequency region is ±7 IRE units (on a voltage scale, an IRE unit is 1 percent of the range from white level to blanking), or ±3 dB.

3.17.2   Envelope delay distortion

Envelope delay is the measure of the time delay of different frequency components as they pass through the system, and is measured in nanoseconds. It is also the derivative of phase with respect to frequency of the phase response of a system. In an ideal system, phase shift is proportional to the frequency, and the delay is constant over the entire spectrum.

As in monochrome, envelope delay causes distortion of the luminance waveform, but in color it can also result in a differential time delay between the subcarrier and luminance signals. The chroma signals in high-chroma areas are then displaced from the luminance (the funny paper effect).

The EIA tolerance for end-to-end chroma transmission time in the vicinity of the subcarrier (3 to 4 MHz) relative to 2 kHz to 600 kHz is ±60 ns.

3.17.3 Differential gain

The transmission system for a color television system must maintain its gain as a function of luminance within close tolerances in the vicinity of the subcarrier frequency. Gain variation at the subcarrier frequency as a function of the luminance signal level is known as differential gain, a form of cross-luminance distortion. Distortion of the chroma/luminance ratio causes a corresponding distortion of the image color saturation.

Tests[3] have shown that differential gain in the range 10 percent to 90 percent of average picture level becomes noticeable to the median observer at ±2 dB and objectionable at ±5 dB. The EIA voltage or current standard for long-haul circuits is ±10 percent or ±0.8 dB.

3.17.4 Differential phase

The eye is even more sensitive to differential phase, another cross-luminance distortion, than to differential gain, since it affects the hue of the image. In the early days of color television, the tolerance for differential phase distortion was generally considered to be ±10°. This tolerance was established more on the basis of the state-of-the-art capability of the then-current transmission systems than on the basis of the visibility of the distortion. Reducing the visibility of the effects of differential phase distortion was perhaps the primary motivation for the development of the PAL system.

With the passage of time and improvements in technology, much tighter tolerances have become practical, and the most recent EIA standard for an end-to-end transmission circuit is ±3°.

3.18 Subcarrier Crosstalk Reduction

The presence of the subcarrier on the luminance channel has the potential of creating crosstalk in the form of annoying dot patterns on the image. This is called cross-luminance distortion. A number of procedures have been developed to minimize the visibility of these patterns.

3.18.1 Relating subcarrier and line frequencies

An important technique that is employed in all subcarrier systems is to relate the subcarrier and line frequencies as described earlier (Secs. 3.14.2 and 3.15.2). This relationship causes the polarity of the dot patterns to reverse or partially reverse on successive scans so that there is a cancellation of their visual effect. It can also cause the dot pattern to be stationary.

When a monochrome receiver is used with a color signal, the line and subcarrier frequencies are no longer locked together, there is no cancellation of the subcarrier on consecutive scans, and a highly visible moving dot pattern can result. This effect is often minimized by the poor aperture response of inexpensive monochrome receivers.

3.18.2 Low-pass filters

A brute force reduction of dot visibility used in early color receivers was a low-pass filter in the luminance channel with a cutoff below the subcarrier frequency—3.56 MHz in the case of NTSC. This limited the horizontal resolution and definition, but the high-frequency output of the image orthicon camera tubes used in this period was so low that very little information was removed by the filter.

3.18.3 Comb filters

The signals from photoconductive camera tubes and CCDs (see Chap. 5) have a sufficiently low noise level to permit aperture correction, and the loss of frequency components above 3.56 MHz significantly degrades the picture. To solve this problem, comb filters were introduced in the receiver luminance channels (see Chap. 13). As their name suggests, comb filters remove alternate bands of frequency components. In color television applications, they remove the chrominance channel frequency components, which are interleaved with the luminance channel components at harmonics of the line frequency (Sec. 3.14.2). Comb filters also have the desirable effect of improving the signal-to-noise ratio by nearly 3 dB, since they remove the noise from these regions of the frequency spectrum.

Although comb filters are highly effective, they do not completely eliminate the problem of subcarrier visibility. The clusters of chrominance sidebands spread when reproducing edges with sharp color differences, so that they are not completely removed by comb filters. This creates artifacts in the form of a dot pattern along the edges.

A similar spreading of the chrominance sidebands occurs if there is rapid motion in the picture or if there are major signal changes from line to line. The luminance channel is also affected because the sidebands surrounding the line frequency harmonics spread into the areas attenuated by the filters.

If the dot pattern on a tricolor kinescope has approximately the same spacing as the subcarrier dot pattern, a heterodyning effect can occur that causes artifacts in the form of a coarse spurious pattern in the picture.

3.19   Component Signal Formats

Component signals are used where it is especially important to avoid the artifacts and cross-color and cross-luminance distortions that are characteristic of composite systems. They are used in complex program production systems where the signals are subjected to repeated recording and rerecording. In such multigeneration systems, it is essential that the distortions in each generation be as low as possible. The requirement for minimum distortion in each generation also encourages the use of digital recording. Digital component formats for recording are described in Chap. 6.

Conventional component systems require three channels (see Sec. 3.6), the Y or luminance channel and two color difference channels. The MAC family (Sec. 3.20) has been developed to transmit, record, and store component signals on a single channel.

3.20   MAC (Multiple Analog Component) Systems

MAC (multiple analog components) is a term describing a family of time-division multiplex signal formats in which the luminance and chrominance components are time-compressed and transmitted sequentially. This avoids the problems of crosstalk and sensitivity to nonlinearity described in Secs. 3.17 and 3.18 for subcarrier systems. These formats also have a better signal-to-noise ratio because of the absence of strong frequency components in the vicinity of the color subcarrier. Synchronizing information can be sent in digital form, thus eliminating the large-amplitude conventional and digital audio signals inserted for a portion of the sync intervals.

At the same time, MAC shares the feature of composite systems of including the entire signal in a single channel.

MAC's disadvantages are its lack of compatibility with broadcast systems and somewhat greater bandwidth requirements. Its principal applications, therefore, are in situations where compatibility is not required and where adequate bandwidth is available. It is particularly useful for satellite transmissions.

3.20.1 The principle of MAC transmission

The principle of MAC transmission is illustrated in Fig. 3.13 by the time relationships in one line of B-MAC, a principal member of the MAC family (see Sec. 3.19.2).

The 63.5-μs line scan time (it is virtually identical for NTSC and PAL) is divided into three segments: 35.0 μs for the luminance signal E_Y', 17.5 μs for the color difference signal $E_R' - E_Y'$ or $E_B' - E_Y'$, and the remainder for blanking, burst, audio, and data (optional). (Alternatively, I and Q signals can be used for transmission of the chroma information.) This requires time compression of all video signal components—by 3:2 for the luminance signal and 3:1 for the color differences. The $E_R' - E_Y'$ and $E_B' - E_Y'$ components are transmitted on alternate lines. As a result, four fields are required to transmit a complete picture, and its vertical chrominance resolution is reduced by one-half.

To achieve the time-compressed format, the luminance component is sampled at a rate of 13.5 MHz, high enough to avoid aliasing, and the color difference components are sampled at one-half this rate, or 6.75 MHz. The samples are stored in memories, and the MAC signal is generated by reading out the samples at a rate of 20.25 MHz in the sequence Y', $R' - Y'$, Y', $B' - Y'$. The luminance samples for one line then occupy 35 μs for a time compression of 3:2, and the color difference samples 17.5 μs, for a time compression of 3:1. The sequence of the samples for a single line is shown in Fig. 3.13. Note that space is reserved for transmission of the burst, audio, and data service.

The process is reversed at the receiver. The samples in the segments

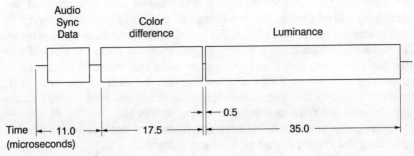

Figure 3.13 MAC transmission.

of each line corresponding to Y', $R' - Y'$, and $B' - Y'$ are stored in a memory bank and read out in real time to generate the luminance and color difference signals simultaneously. They can then be matrixed and processed to recover the original E'_R, E'_G, and E'_B.

The resolution of the chrominance signal is reduced both horizontally and vertically by this process—horizontally because the time compression is twice as great, and vertically because only alternate lines are transmitted.

To achieve resolution of the luminance component that is equivalent to that of the uncompressed signal, the bandwidth requirement is increased by an amount that is inversely proportional to the time compression. Thus, to achieve resolution equivalent to that of an uncompressed 4.2-MHz luminance component, a bandwidth of 6.3 MHz (4.2×1.5) is required for the compressed signal.

3.20.2 MAC formats

A number of MAC formats have been developed to meet the needs of various applications. The complete family includes ACLE, A-MAC, B-MAC, C-MAC, D-MAC, HD-MAC, HD-MAC60, MAC-60, and MUSE. The formats for the video portions of the signals are identical in many cases; the differences are in the audio and data channels.

B-MAC, described above, is the most common. It requires a bandwidth slightly in excess of 6 MHz (as compared with 4.2 MHz for an NTSC signal).

C-MAC is an occasional variation. It requires a bandwidth of 20 MHz, the additional bandwidth being used for data.

HD-MAC has been proposed for high-definition television in Europe. It operates at 50 fields per second, with 1250 scan lines interlaced per frame. The aspect ratio is 16:9.

MUSE (multiple sub-Nyquist sampling encoding) was developed in Japan for the transmission of HDTV signals by satellite; see Chap. 14.

References

1. K. Blair Benson, (ed.), *Television Engineering Handbook,* McGraw-Hill, New York, 1986.
2. EIA/TIA Standard, "Electrical Performance for Television Transmission Systems," EIA/TIA-250-C, Washington, D.C., 1990. FCC Rules, Part 73. CCIR, *Recommendations and Reports, CCIR, 15th Plenary Assy.,* Vol. XI, Broadcasting Service (Television), ITU, Geneva, 1982.
3. D. G. Fink, *Television Engineering Handbook,* McGraw-Hill, New York, 1957.

Digital Television

4.1 Introduction

The use of the digital format for the transmission, recording, and processing of video signals is comparatively recent, and in a few short years it has revolutionized many aspects of television technology. For some applications, the digital mode has replaced analog because it can perform functions that are difficult or impossible in a conventional analog format. (Conversely, some functions, such as filtering, can be performed more easily in the analog format.)

Most television technologies were developed solely for use in that industry. Digital technology is an exception. By the time its value for video systems was recognized, it had reached a high degree of development for use in data processing and communication systems. As a result, digital television is heir to a vast body of technical literature that encompasses many volumes and that grows exponentially year by year.

A complete treatment of digital technology as employed in video systems is beyond the scope of this chapter. Rather, this is a summary of its fundamentals and a foundation for the more complete treatments found in the references. It also provides an introduction to Chaps. 6, 8, and 14, which describe video recording, editing, and HDTV.

4.2 Analog and Digital Formats

Conventional video signals, as described in Chaps. 1 to 3, are the electrical analogs of scene brightness. They vary continuously and are described as having an analog format. At the receiver, the eye responds to analog displays with continuous brightness variations. Video signals, therefore, are inherently analog, both when generated and when displayed. Between these endpoints, however, there is the option of converting the signal to a digital format.

The essential steps in the conversion of a video signal from an analog to a digital format, usually described as an A-to-D conversion, are shown in Fig. 4.1.

The analog signal is sampled in the time domain at uniformly spaced sampling points. The amplitudes of the samples are defined by discrete, preestablished quantized levels, which are identified for each sampling point. Each level is then described by an encoded "word" or byte (the encoding in Fig. 4.1 is offset binary; see Sec. 4.11.3), which is a series of "on" or "off" pulses or bits that can be processed, recorded, and transmitted. Since the bits in the digital signal have only two possible states, this is known as a binary system. The number of quantized levels that can be described is 2^n, where n is the number of bits per word.

The choice of number of bits per word and thus the number of levels that can be identified, involves an important tradeoff in the design of a digital system. If there are too few, the picture will have a blocky appearance. On the other hand, the choice of too many unnecessarily increases the bandwidth requirement. A common choice for television recording and transmission systems is 8-bit words that can identify 2^8 or 256 different signal levels. In practice, the number of levels that are identified is somewhat lower, typically 235, to allow for temporary overloads; see Sec. 4.9.1.

For some applications, 10-bit words with the capability of identifying 1024 levels are used.

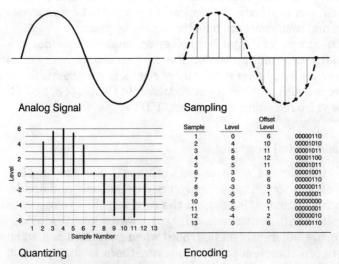

Figure 4.1   Analog-to-digital conversion.

At the receiver, a reverse D-to-A conversion takes place, and a reasonable facsimile of the original analog signal is recovered for display.

4.3 Comparison of Digital and Analog Signal Formats

Digital signals have important advantages and disadvantages compared with analog.

4.3.1 Bandwidth requirements

The most obvious difference between digital and analog signals is their bandwidth requirements.

Digital signals inherently require more bandwidth. A commonly used format for composite signals employs a sampling rate that is 4 times the subcarrier frequency, or 14.4 MHz (NTSC). With an 8-bit word length, the bit rate is 115.2 Mbits/s, and the bandwidth required is approximately one-half this, or about 58 MHz. This compares with 4.25 MHz for the analog signal. If bits are added to each word for error correction (see Sec. 4.15), an even greater bandwidth will be required.

In practice, however, this additional bandwidth is often not required for video signals; on the contrary, the digital format has the potential for enormous bandwidth reduction. Systems have been proposed that compress the bandwidth in the ratio of 100 to 1 or more (although this degree of compression would probably have some adverse effect on image quality).

Bit rate and hence bandwidth reduction is achieved by taking advantage of the unique characteristics of video signals, such as their repetitive and predictive nature. The capability for bandwidth reduction has the potential to have an enormous impact on the television industry.

4.3.2 Signal-to-noise ratio

One of the most important advantages of the digital format is its relative invulnerability to the introduction of noise in the transmission and recording processes.

Noise power is additive in analog systems, and the system signal-to-noise ratio is the sum of the noise contributions from each stage. Thus the signal-to-noise ratio for the system is inevitably lower than that of the worst stage.

With the digital format, noise causes *bit errors*—for example, an "on" pulse is transformed to an "off"—but so long as the *bit error rate* (ber) is low enough, *error correction* circuits can restore the original bit stream. The effect of even larger errors can be minimized by *error concealment* circuits. There is little or no reduction of the signal-to-noise ratio as noise power is increased until the bit error rate becomes so large that error correction or error concealment is not effective and the signal is lost. By contrast, the signal-to-noise ratio in analog systems suffers from *graceful degradation*. Its quality gets steadily worse, although it remains useful, as noise is added.

This property of digital systems is particularly useful for program production with complex editing functions and multiple-generation recordings. For this purpose, digital recorders are almost universally used. This capability is equally useful in multihop transmission systems, where each leg of the circuit adds noise and distortion to the signal. Figure 4.2 shows the comparative performance of analog and digital recorders in a multigeneration application.

Digital transmission has one disadvantage for broadcast use, however. With the graceful degradation of analog transmission, a usable, albeit poor, signal-to-noise ratio may be encountered at points on the fringe of the service area where digital signals are no longer usable.

4.3.3 Nonlinear distortion

As with noise, the digital format is relatively immune to distortions caused by the nonlinear transfer functions of transmission or recording circuits. Also as with noise, this property is an important advantage with multigeneration systems. Its value is particularly great

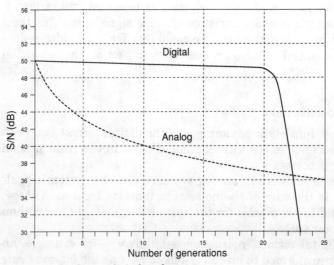

Figure 4.2 Analog vs. digital performance.

with the NTSC color system, with its sensitivity to differential gain and differential phase distortion.

4.3.4 Aliasing

A digital signal results from sampling in the vertical dimension by the scanning lines and in the horizontal dimension by the sampling intervals. As a result, the digital signal has the potential for aliasing both vertically and horizontally by the effect described in Sec. 2.17.6. The magnitude of vertical aliasing is the same for analog and digital systems. The magnitude of horizontal aliasing depends on the amplitude of the frequency components that exceed the Nyquist limit for the sampling rate. It can be avoided by using a sampling rate that is at least twice the highest-frequency component in the analog signal.

4.3.5 Cost and complexity

Digital circuitry is inherently more complex than analog, and it was considerably more expensive when it was first introduced. In addition, the planning, operation, and maintenance of digital equipment has been foreign to the experience of most television engineers. All of these problems have been eased by the rapid growth of the digital communication and computer industries. This has encouraged the growth of a cadre of engineers and technicians trained in digital techniques. And they have stimulated the development of large-scale and very large-scale (VLSI) integrated circuits that have greatly reduced the cost of digital equipment. As a result, many digital systems are now actually cheaper than their analog counterparts.

4.3.6 Signal processing

The digital format is capable of performing signal conversion and processing functions that would be difficult or impossible with analog signals. After an A-to-D conversion, the signal consists simply of a series of numbers, "on" and "off" bits, that can be manipulated with little or no loss of picture quality.

This capability is enhanced by the ease with which the bit signals can be stored in memories for rapid retrieval as needed. Memory technology has developed rapidly in recent years in response to the needs of the computer industry, and much of it is applicable to digital video signals.

Four important signal processing functions that can be performed with digital signals are time-base error correction (Sec. 4.5.1), standards conversion, e.g., PAL to NTSC (Sec. 4.5.2), post-production editing (Chap. 8), and bandwidth reduction (Sec. 4.18).

4.3.7   Co-channel TV station spacing

The digital format has the potential, as yet (1991) unproven, of enabling co-channel TV stations to operate without excessive interference at much closer spacing than is possible with analog signals. This is partly because digital baseband signals are less vulnerable to co-channel interference and partly because the major source of interference in analog transmissions is their sync and blanking pulses (see Sec. 1.8), which can be replaced by digital code words.

The co-channel spacings now required by the FCC (see Chap. 9) vary from 155 to 220 miles, depending on the geographic location and frequency band. Rough calculations indicate that these might be cut nearly in half with digital baseband signals. This would increase the number of stations that could operate on the same channel by a factor of nearly 4.

Reduced spacing of co-channel stations, combined with the bandwidth reduction of the digital format (see Sec. 4.3.1), has the potential of enabling as many TV stations to broadcast HDTV signals as now broadcast NTSC or PAL signals.

4.3.8   Multipath transmission (ghosts)

The resistance of digital signals to co-channel interference would also reduce the problem of ghosting as a result of multipath transmission in television broadcasting.

4.4   Transmission and Recording Formats

4.4.1   Digital modulation modes

All digital modulation modes establish two "on-off" conditions of amplitude, frequency, or phase, or of a combination of two of these parameters. There is a wide variety of possibilities, and a summary comparison of the speed vs. bandwidth relationship and the power requirements of four modes that span the range of possibilities is shown in Table 4.1. These modes are *on-off keying* or *OOK*, which is an elementary form of *amplitude-shift keying (ASK)*, *quaternary phase-shift keying (QPSK)*, *frequency-shift keying (FSK)*, and *quadrature amplitude modulation (QAM)*.

Among these four modes, QAM can transmit the highest bit rate in a fixed bandwidth, but the power level must be high to achieve a satisfactory ber. Accordingly, it is often used in microwave systems where adequate power is available. QPSK presents an attractive compromise between power requirements and transmission speed, and it

TABLE 4.1   Performance of Digital Signal Transmission Modes

	OOK	QPSK	FSK	QAM
Transmission speed, bits/Hz	0.8	1.9	0.8	3.1
Required $Eb\backslash Ea$ for BER = 10^{-4}	12.5	9.9	11.8	13.4

Key: E_b = energy per bit
E_a = white noise spectral density
OOK = on-off keying
QPSK = quaternary phase-shift keying
FSK = frequency-shift keying
QAM = quadrature amplitude modulation
SOURCES: J. D. Oetting, "A Comparison of Modulation Techniques for Digital Radio," *IEEE Transactions on Communications,* 23(12) 1979; Alauddin Javed, "Signal Transmission Modes," in A. F. Inglis (ed.), *Electronic Communications Handbook,* Chap. 11, McGraw-Hill, New York, 1988.

is frequently used in satellite systems. The references in Table 4.1 describe other digital transmission modes.

4.4.2 Serial and parallel transmission

The bits in a character code can be transmitted or recorded either serially over a single path or in parallel with a separate path for each numbered bit, such as the first bit in each word. Thus the transmission (or recording) of an 8-bit word by parallel transmission requires eight separate communication paths. The use of parallel transmission simplifies the terminal equipment, but the necessity for a multi-conductor transmission path makes it impractical except for short distances, such as the interconnections within a studio. Serial transmission, therefore, is more commonly used for medium and long distance transmission.

4.4.3 Component and composite signals

Either component or composite color signals (see Chap. 3) can be converted to a digital format.

The limitations of composite color signals are especially serious for post-production editing techniques that require extensive use of multigeneration recording. Component signals are supportive of the advantages of the digital format—lower noise, lower distortion, greater flexibility for editing, and freedom from artifacts from luminance-subcarrier crosstalk. Accordingly, the component and digital formats are frequently used in combination, particularly in recording. See Chap. 6.

In many system applications, however, it is necessary or desirable to transmit or record the signal in a composite form, such as NTSC or

TABLE 4.2   Sampling Frequencies

Designations:	NTSC, SMPTE D-2		PAL, SMPTE D-2
	$3f_{sc}$	$4f_{sc}$	$4f_{sc}$
Composite Signals			
Bandwidth, MHz	4.2	4.2	5.5
Subcarrier frequency (f_{sc}), MHz	3.58	3.58	4.43
Sampling frequency, MHz	10.6	14.4	17.7
Samples per total line	674	915	1132
Samples per active line	557	757	939
Bit rate, Mbits/s (8-bit words)	85.9	114.5	141.9
Component Signals			
	525 lines, 59.95 fields		625 lines, 50 fields
Designations:	SMPTE D-1 SMPTE RP 125 CCIR 601 4-2-2		SMPTE D-1 EBU 3246/3247 CCIR 601 4-2-2
Luminance channel			
Bandwidth, MHz	5.5		5.5
Sampling frequency, MHz	13.5		13.5
Samples per total line	858		864
Samples/active line	710		716
Bit rate, Mbits/s	108.0		108.0
Color difference channels			
Bandwidth, MHz	2.2		2.2
Sampling frequency, MHz	6.75		6.75
Samples per total line	429		432
Samples per active line	355		358
Bit rate, Mbits/s (8-bit words)	54.0		54.0

PAL. As a result, industry standards have been established for the frequency and location of sampling points for composite and component signals. See Sec. 4.7 and Table 4.2.

4.5   Signal Processing Functions

This section describes some of the more important signal processing functions that are facilitated or made possible by the digital format.

4.5.1   Time-base error correction

Maintaining a stable time base for video signals and accurately preserving their original time relationships is a problem with videotape recorders because of the timing errors introduced by mechanical recording systems. The time-base errors at the output of uncorrected helical scan recorders, with their long recording paths on an elastic tape medium, can be particularly severe.

An adequate degree of stability for consumer-type recorders can be achieved by the "color under" technique (see Chap. 7), but more precise correction is required for professional recorders because of the necessity of performing editing and other multigeneration functions on their outputs. A number of methods have been developed for providing precise time-base correction with the signal in analog form, all of them using a variable electrical delay line as their basic element. Timing errors of less than ±2 ns, approximately ±2° at the subcarrier frequency, have been achieved. This a satisfactory tolerance, but the amount of correction required for helical scan recorders is so great that long and expensive delay lines, having a range of several microseconds, are required.

Time-base correction can be performed more cheaply and easily with digital time-base correctors. The availability of low-cost correctors has led to the nearly complete replacement of quadruplex videotape recorders by helical scan; see Chap. 6. The improvement in stability that can be achieved is quite impressive. Timing errors as great as 10 μs or greater can be reduced to a small fraction of a microsecond.

The principle of operation of digital time-base correctors is straightforward. The bit stream at the output of the recording process is read into the memory in real time. It is then read out at a rate that is synchronized accurately with the system's subcarrier. In effect, the memory acts as a buffer that absorbs the timing errors in its input signal and reads out a time-corrected signal.

4.5.2 Standards conversion

Prior to the advent of digital television, standards conversion was accomplished by the brute force methods of kinescope recording or of focusing a camera operating at the new standards at an image on a kinescope operating at the old. The picture quality of analog conversion systems was marginal at best. The standards-converted images suffered not only from the degradations that are inherent in the conversion of line and frame rates, but, much worse, from the cumulative degradation of the signal-to-noise ratio, gray scale, and definition that are unavoidable in multigeneration analog systems.

The use of the digital format does not solve the inherent problems of rate conversions, but it eliminates or greatly reduces the cumulative quality degradation of analog systems.

Digital standards converters employ the same principle as time-base correctors—storage at the old standards and read-out at the new.

The process is considerably more complex, however, because the number of lines, fields, and frames must be changed without destroying the time relationships of the events in the image. Also, it is difficult to convert the burst and subcarrier from one standard to another, for example, from NTSC to PAL, and the first step in the conversion process is to convert the composite signal to its luminance and color difference components. The line, field, and frame rates are then converted to the new standards, and the signal components at the new standards are reconverted to the new composite format.

The signal to be converted is read into a *frame memory* that stores a complete frame at the original standards. The signal is then read out at the new standards.

The difference between the line period for 25-Hz, 625-line systems and 30-Hz, 525-line systems (64.0 vs. 63.6 μs) is so small that it is usually ignored. The change in the number of lines per frame, from 525 to 625, and in the number of frames per second, however, must be taken into account in the conversion process. Repeating or deleting scanning lines or frames is not a satisfactory solution. Repeating or adding scanning lines would cause discontinuities in the image. Repeating or deleting fields or frames would cause an appearance of jerky motion. The solution is to generate signals by interpolation that approximate what the new signals would have been had they been originally produced at the new standards.

4.5.3 Post-production editing

The art of post-production film editing—the production of an edited master from a combination of short film or tape segments and special effects—was developed over a period of decades by the motion picture industry, and its practitioners are highly skilled artisans. The television industry tried for years to duplicate the complex post-production functions performed routinely in film studios, but these efforts were only modestly successful with analog systems.

The use of the digital format, which can produce multiple generations of recorded material without significant loss of quality, incorporate precise computer control for switching and effects, and insert computer-generated program material—all while the results are being monitored—has made television an extraordinarily powerful post-production medium. With the improvements that have occurred in the performance of television cameras, television production techniques are now practical for the production of films for use on television.

The digitized analog signals used for editing can be either composite, such as NTSC or PAL, or in component form, usually R, G, and B or Y, $R - Y$, and $B - Y$. Composite digital signals have the same basic problems of luminance-chrominance crosstalk as analog, and it is desirable, therefore, to perform complex editing functions with component signals.

Television editing systems are described in Chap. 8.

4.5.4 Bandwidth reduction systems

It is an apparent paradox that digital formats, which inherently require more bandwidth than analog, can be used to reduce the bandwidth required for video signals. The technologies for reducing the bandwidth of video signals by digital techniques are developing rapidly, and it has been estimated that as many as 40 companies worldwide are engaged in this endeavor. Bandwidth compression can accommodate the growing requirements of satellites, cable systems, and HDTV. Current digital bandwidth reduction systems are summarized in Sec. 4.18.

4.6 Analog-to-Digital Conversion

Most digital video signals are originally generated in analog form, and analog-to-digital (A-to-D) conversions are an integral part of digital systems. Figure 4.3 shows functional diagrams of A-to-D converters for composite and component analog signals.

The first element (optional) is an *antialiasing* filter. This is a low-

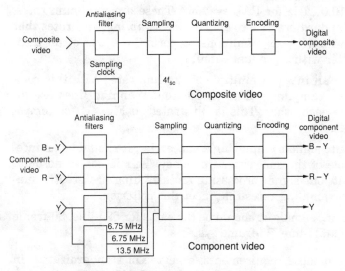

Figure 4.3 Analog-to-digital converters.

pass filter with a cutoff below the Nyquist limit (see Sec. 2.18.1), i.e., below one-half the sampling rate. This eliminates the signal frequency components that could cause aliasing. It is desirable to have the sampling rate high enough that there is a reasonable separation between the nominal cutoff frequency of the filter and the Nyquist frequency. This makes it possible to use a filter with a gradual cutoff, without the excessive phase shift of sharp-cutoff filters. In the component format, the cutoff of the filter in the color difference channels is approximately one-half that in the luminance channels in accordance with their reduced bandwidth.

The antialiasing function in the A-to-D converter is followed by the sampling, quantizing, processing, and encoding functions.

4.7 Sampling

4.7.1 Sampling rates

The sampling rate or its reciprocal, the sampling interval, is a fundamental specification. It is one of the major concerns of national and international standardizing bodies, including the SMPTE, IEEE, EBU, and CCIR.

The sampling rate should meet a number of conditions:

1. It should meet or exceed the Nyquist limit to avoid degrading the horizontal resolution and to minimize aliasing (Sec. 2.17.6). This requires it to be at least twice the upper limit of the analog signal frequency components, or 8.5 MHz (2 × 4.25 MHz) for NTSC systems and 10.0 MHz for PAL systems. These are minimum rates, and it is standard practice to employ *oversampling* with rates that are higher than these minimums to permit the use of an antialiasing filter with a gradual cutoff.

2. It should be an integral multiple of the line rate. With this relationship, the sampling points on adjacent scanning lines are directly above each other. This is illustrated in Fig. 4.4 for component signals.

3. For composite signals employing a subcarrier, it should be an integral multiple of the subcarrier frequency. This is not an absolute requirement, but signal processing and decoding are greatly simplified with sampling rates that meet this criterion.

4. For component signals, it should be the same for 525-line, 30-frame systems and 625-line, 25-frame systems.

In response to these requirements, a degree of standardization in sampling rates has been achieved, as shown in Table 4.2. (D-1 is a des-

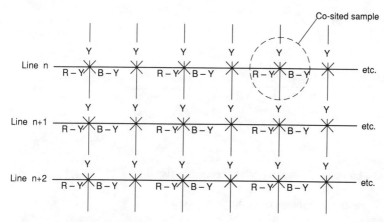

Figure 4.4   Location of sampling points, component signal.

ignation for a standard component recording format and D-2 for a composite. See Chap. 6.)

The sampling rate of 13.5 MHz was chosen for component signals because it is an integral multiple of both the NTSC and PAL line rates. It is not, however, an integral multiple of either the NTSC or PAL subcarrier frequency, and rates of $3f_{sc}$ or $4f_{sc}$ can be used. A rate of $3f_{sc}$ does not exceed the Nyquist frequency by a comfortable margin, and so a sampling frequency of $4f_{sc}$ is the most commonly used and is an SMPTE standard.

4.7.2   Location of sampling points

The locations of the sampling points on individual lines for the D-1 component format are shown in Fig. 4.4. The luminance and two color difference sampling pulses are synchronized so that the color difference points are co-sited with alternate luminance points.

In the D-2 format for composite signals (see Sec 6.14), the sampling points have a specified relationship with the phase of the burst. For both PAL and NTSC, there are four sampling points for each cycle of the burst. For PAL, the sampling points are at the 0°, 90°, 180°, and 270° points of the burst waveform. For NTSC, the first sample is located at 57°, the I axis (Sec. 3.14.1). The others are located at 147° (the Q axis), 237°, and 327°. See Fig. 4.5. This is known as I-Q axis sampling.

The sampling rate and phase must be synchronized with the line and subcarrier frequencies and phases in order to maintain these precise locations. For composite signals, the synchronizing element can be the subcarrier burst. This system element that generates the sampling frequency is called the *sampling clock*.

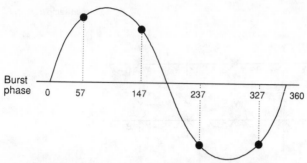

Figure 4.5 Location of sampling points, composite signal.

4.7.3 Aperture response of sampling processes

If the sampling pulses had zero width, the output of the sampling process would be a stream of pulses having amplitudes equal to that of the original analog signal at the sampling points. In practice, the pulses have a finite width or aperture, and the amplitude of the samples will be determined by the amplitude of the analog signal averaged over the pulse widths. The width of the pulses establishes the aperture response of the sampling process; at the sampling rates and widths employed in digital television systems, this effect is small, although it is similar in principle to that produced in the vertical dimension by the scanning process.

Table 4.3 shows the aperture response at 4.2 MHz of the sampling process with a sampling rate of 14.32 MHz and a range of sampling pulse widths varying from 12.5 to 100 percent of the sampling interval. In this example, the signal component frequency, 4.2 MHz, is well below the Nyquist limit for a 14.32-MHz sampling rate. The effect of pulse width on aperture response would be greater for higher-frequency signal components.

TABLE 4.3 Aperture Response at 4.2 MHz of Sampling Process with 14.32-MHz Sampling Rate

Pulse width/sampling interval	Aperture response
0.125	1.0
0.25	1.0
0.5	0.94
1.0	0.80

4.8 Bit Rates and Bandwidth Requirements

The bit rate is equal to the number of bits per word times the sampling rate, and the bandwidth requirement is approximately one-half

TABLE 4.4   Digital Television Bit Rates and Bandwidth

System	Bit rate, Mbits/s	Bandwidth, MHz
NTSC	114.5	58
PAL	141.9	72
Component		
Luminance	108	54
Color difference	54	27

the bit rate. For 8-bit words as used in digital television, these are shown in Table 4.4. These bandwidths are extremely wide, and they often require some form of bandwidth compression.

4.9   Quantizing

The next step in the A-to-D process is to quantize the pulse stream resulting from the sampling process.

4.9.1   Quantized signal levels

Quantizing is depicted in Fig. 4.6. The signal amplitude range is divided into discrete intervals, and a fixed or quantum level is established for each interval. This level is assigned to samples having an analog value that falls within the interval. The difference between the quantum level and the analog signal at the sampling point is the *quantizing error*.

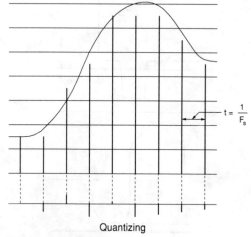

$$t = \frac{1}{F_s}$$

Quantizing

Figure 4.6   Quantizing.

As noted in Sec. 4.2, most television digital quantization systems are based on 8-bit words with 2^8 or 256 discrete values (2^{10} or 1024 levels are sometimes used for pre-gamma signals). Something less than 256 levels are identified, because of the need to use some of the values for audio, data, and control signals. Figure 4.7 shows typical utilization of the 256 quantized levels for composite and component signals. The entire composite signal, including sync pulses, is transmitted within the quantized range of levels 4 to 200.

In component systems, sync pulses fall outside the quantized range and are not transmitted. They are unnecessary because line and field synchronization are provided by coded pulses. The luminance signal occupies levels 16 to 235. The color difference signals can be either positive or negative, and the zero black level is placed at the center of the quantized range because there are no negative numbers in the binary number system used for encoding quantized signals.

Note that neither range includes the binary words 00000000 (0_{10}) or 11111111 (255_{10}) (see Sec. 4.10); they are available for synchronizing. (The subscript 10 means that the number is expressed to the base 10; binary numbers could be designated by the subscript 2.)

4.9.2 Quantizing errors

If the sampling rate is harmonically related to a signal frequency component, the result of quantizing errors is distortion of the waveform.

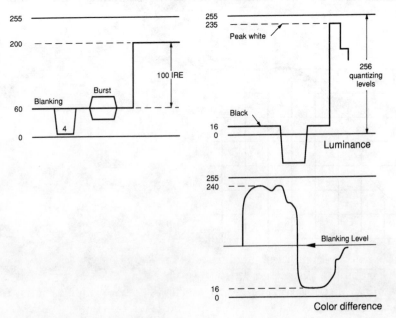

Figure 4.7 Quantizing level utilization.

In the more unlikely event that the relationship is random, the result of quantizing errors is noise. The peak-to-peak signal-to-rms quantizing error ratio is given by Eq. (4.1):

$$\frac{S}{Q_e} \text{(dB)} = 10.8 + 6.02n + 10 \log \left(\frac{F_s}{F_v}\right) \tag{4.1}$$

where S = peak-to-peak signal
Q_e = rms quantizing error
n = number of quantum levels
F_s = sampling rate
F_v = video bandwidth

For $n = 8$, $F_s = 14.4$ MHz, and $F_v = 4.2$ MHz, $S/Q_e = 64.3$ dB.

The nature of the noise depends on the magnitude of n. If n is 6 or greater, the noise will be random, as from thermal effects. If n is less than 6, it will appear as false contours on sharp edges.

Equation (4.1) assumes that the original analog signal is noise-free. With a noise-free input signal, the sampled waveform will be distorted when it is reconverted to an analog signal. For example, a diagonal line in the image will be reproduced as a stairstep, with discontinuities equal to the quantum intervals. If there is noise in the analog signal, however, it will partially or totally obscure the discontinuities. If the rms value of the noise exceeds one-third the quantum interval, the discontinuities will disappear in the noise. The ratio S/Q_e under these conditions is given by Eq. (4.2):

$$\frac{S}{Q_e} \text{(dB)} = 6.02n \tag{4.2}$$

When $n = 8$, $S/Q_e = 48.2$ dB.

In some cases it may be desirable to trade off signal distortion for noise by deliberately introducing noise. This is *dithering*, an intentional rapid and random variation of some parameter of a signal that eliminates the visibility of quantum discontinuities.

4.10   Encoding

The final step in analog-to-digital conversion is to encode the quantized levels of the signal samples.

The encoding and processing of digital signals has been the subject of intensive mathematical analysis. Hundreds of papers have been published on the subject, as well as a number of excellent textbooks. In addition, the literature is growing at a phenomenal rate as industrial laboratories in all the developed countries continue to explore new applications for the digital format. A complete description of this subject, therefore, is beyond the scope of this chapter. Rather, the purpose here is to describe the basics and define the principal terms of digital signal encoding as a foundation for a more extensive study of the subject.

4.11   Binary Numbers

Most encoded digital signals utilize *binary* numbers, which have 2 as their base as contrasted with the base 10 of conventional decimal numbers. They are used for digital signals because they naturally describe the operation of systems that have only two states, "on" and "off."

Several binary notations are in use to meet the requirements of different applications. Five of them are described here—pure binary, octal, hexadecimal, offset binary, and two's complement.

4.11.1   The pure binary code

Each character in a binary system, called a *bit,* is either "1" or "0." A group of bits, known as a *word* or a *byte,* represents a decimal number. The group can be separated by a *radix point,* the binary equivalent of a decimal point. The first bit to the left of the radix point has the value 1, the next to the left 2, the next 4, and so forth. The first bit to the right of the radix point has the value ½, the second ¼, and so forth. The decimal equivalent of the word is the sum of these values. This is illustrated in Fig. 4.8, which describes the representation of a decimal number in the pure binary code. The subscript following a number indicates whether it is to the base 10, as in a decimal system, or to the base 2.

The significance of a bit in a digital word varies greatly depending on its location in the word. In a pure binary system, the least significant bit (LSB) is on the right, while the most significant bit (MSB) is on the left of the word. In an 8-bit word, an error in the LSB leads to an error of only 1 in 256—probably not noticeable in the image. An

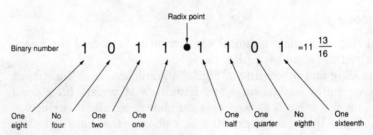

Figure 4.8   Binary and decimal numbers.

error in the MSB would cause an error of 128 quantum levels, more than half the amplitude range of the signal. This property is extremely important in the design of error correction and other signal processing circuits.

Placing the MSB on the left is an arbitrary convention, and in some cases the sequence is reversed so that the LSB is on the left.

4.11.2 Octal and hexadecimal codes

Words in the pure binary code become extremely long when they express large numbers. The number 255_{10}, for example, becomes 01111111_2. To express these numbers with shorter words, hybrid binary/decimal codes, *octal* and *hexadecimal*, have been developed. In the hexadecimal system, the binary numbers are defined in groups of four, starting at the LSB, with the letters A to F assigned for numbers above 9. In the octal system, the binary numbers are defined in groups of three and assigned decimal numbers in sequence (8 and 9 are omitted). Table 4.5 compares the three codes for a range of decimal numbers.

As noted previously, most digital television systems use 8-bit words that can identify 2^8 or 256 values. This is usually adequate for the video signal itself, but it is sometimes desirable to go to 10-bit words for the transmission of data or auxiliary functions. Ten-bit words can identify 1024 values.

TABLE 4.5 Binary Codes

Decimal	Pure binary	Hexadecimal	Octal
001	00000001	1	01
002	00000010	2	02
003	00000011	3	03
004	00000100	4	04
005	00000101	5	05
006	00000110	6	06
007	00000111	7	07
008	00001000	8	10
009	00001001	9	11
010	00001010	A	12
011	00001011	B	13
012	00001100	C	14
013	00001101	D	15
014	00001110	E	16
015	00001111	F	17
020	00010100	14	24
100	01100100	64	144
200	11001000	C8	310
255	11111111	FF	377
256	100000000	100	1000

4.11.3 Offset binary code

The *offset binary* is a code that can express negative numbers in a digital format. This is achieved by adding a fixed amount to the analog signal before it is sampled and quantized so that the sum is always positive. In the CCIR-601 standard, for example, the blanking level for the analog color difference signals is set at the quantum level 128_{10}, and signal excursion is from 16_{10} to 240_{10}. The corresponding binary numbers are 00010000 and 11110000 in the offset binary code. See Fig 4.7.

The offset binary code is also used for composite signals and to avoid clipping.

4.11.4 Two's complement code

The *two's complement code,* which also has the capability of representing negative numbers, is frequently used for processing functions such as signal mixing. It is produced by defining the words in the upper half of the word's number range as negative numbers.

At the first binary number for which the MSB is 1, the sign of the corresponding decimal number is reversed, and the absolute value of decimal numbers counts down as illustrated in Table 4.6 for 4- and 5-bit words.

The largest positive decimal number that can be represented in the

TABLE 4.6 Two's Complement Code

| | 4-bit words | | | 5-bit words | | |
| | Decimal number | | | Decimal number | |
Binary word	Pure binary	Two's complement	Binary word	Pure binary	Two's complement
0001	1	1	01001	9	9
0010	2	2	01010	10	10
0011	3	3	01011	11	11
0100	4	4	01100	12	12
0101	5	5	01101	13	13
0110	6	6	01110	14	14
0111	7	7	01111	15	15
1000	8	− 8	10000	16	− 16
1001	9	− 7	10001	17	− 15
1010	10	− 6	10010	18	− 14
1011	11	− 5	10011	19	− 13
1100	12	− 4	10100	20	− 12
1101	13	− 3	10101	21	− 11
1110	14	− 2	10110	22	− 10
1111	15	− 1	10111	23	− 9

two's complement code is $2^{n-1} - 1$. The absolute value of the largest negative number that can be represented is 2^{n-1}. The two's complement code for positive numbers equal to or less than 2^{n-1} is the same as its pure binary code. The two's complement code for a smaller negative number, $-X$, is equal to the pure binary code for the positive number $2^n - |X|$.

The first digit in this code for positive numbers is always 0, while for negative numbers it is 1.

4.12   Synchronization

The digital bit stream from the A-to-D converter must convey not only analog signal information but also digital words to synchronize it with the D-to-A converter at the output of the system. This is accomplished by the use of digital code words, which make it possible for the D-to-A converter to identify positively the following:

1. In serial systems, the beginning of each word.

2. In component systems, the location of the beginnings of lines, the active regions of lines, fields, and frames.

The code words must have the characteristic that the decoder does not confuse them with textual words. Bits 00000000 and 11111111, which are outside the quantizing range for the text (see Sec. 4.9.1) meet this requirement.

Figure 4.9 shows a four-word synchronization pattern, known as TRS-ID, that is used for serial transmission. It is repeated for each line at the end of the horizontal sync pulse and synchronizes word beginnings, lines, fields, and frames. The first three words, the TRS, define the LSBs and MSBs in the bit stream. The fourth word, the ID, contains the bits that identify odd and even fields, the beginning and

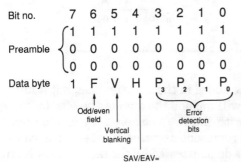

Figure 4.9 TRS-ID serial synchronization pattern.

end of the active line time, and the location of vertical blanking. Since each bit in the ID has only two states, the identification is limited to two choices, such as the beginning and end of the active line time. In some cases, the ID has bits that are used for parity checking and error correction.

4.13   Binary Arithmetic

A branch of mathematics known as *Boolean algebra* (for the English mathematician George Boole), which deals with the relationships between binary numbers, has been developed. It includes a series of laws on deductive logic that can be used for the analysis of complex relationships between many interacting components, for example, in the design of machine control systems. It is commonly used for the addition and multiplication of binary signals.

4.13.1   Binary addition

In principle, the addition of pure binary numbers is similar to that of decimal numbers. Starting with the LSB column, the binary digits in each column are added. If the sum of the digits, n, in a column is 0 or 1, that number is entered as the total for that column. If n exceeds 1 and is an even number, 0 is entered as the total and $n/2$ is carried into the next column to the left. If n is an odd number, 1 is entered in the total and $(n - 1)/2$ is carried into the next column to the left. Two examples are:

Adding Binary Numbers

	Binary number	Decimal equivalent		Binary number	Decimal equivalent
Carryover	1 1 1			1 1	
Binary A	1 1 0 1	13		1 0 1	5
Binary B	1 1 1 1	15		1 1 1	7
Binary C	0 0 1 1	3	Total	1 1 0 0	12
Total	1 1 1 1 1	31			

Offset binary numbers can be added in the same fashion, but it must be remembered that the sum will include the sum of the offsets.

Addition of binary signals with the two's complement is often necessary because of its ability to represent negative numbers. The value of the two's complement code is demonstrated by the fact that addition can be carried out using the same rules as with pure binary numbers.

Adding Binary Numbers, Two's Complement Code

	Binary number	Decimal equivalent
Carryover	1 0 1	
Binary A	0 1 1 1	7
Binary B	0 1 0 0	4
Binary C	1 0 1 1	−5
Total	0 1 1 0	6

When adding positive numbers, whether using pure binary or two's complement code, care must be taken that the sum does not exceed the range available for the system, i.e., 2^n for the pure binary code and $2^{n/2}$ for the two's complement. If the overload results from negative numbers in the two's complement format, some bits may actually be deleted. Some systems are provided with overflow control circuits to avoid this effect.

4.13.2 Binary multiplication

Binary multiplication is not as simple as addition, but it is essential to many signal processing functions, such as fading and mixing.

Multiplication can be accomplished by repeatedly adding the bits in each multiplicand column in a shift register in accordance with the values in the multiplier. The sum for each column is weighted in accordance with its significance, ranging from the LSB to the MSB.

Binary Multiplication

Binary notation	Decimal numbers
[1 0 1]	5 [4 + 0 + 1]
× [1 1 1]	7 × [4 + 2 + 1]
= [1 0 0 0 1 1]	7 × 5 = [16 + 4]
	+ [8 + 2]
	+ [4 + 1] = 35

This procedure requires a tremendous number of operations, but they can be accomplished quickly and accurately with integrated circuits.

4.14   The Transmission of Digital Signals

As noted in Sec. 4.4.2, encoded signals can be transmitted serially or in parallel. In parallel transmission, a separate circuit is provided for each of 8 (or 10) bits. In serial transmission, the 8 (or 10) bits in each word, which are produced simultaneously, are stored in a *shift register* and read out sequentially to a single circuit.

The minimum bandwidth requirement of serial channels is nominally one-half the bit rate (see Table 4.4). An eight-channel parallel system would require eight separate channels, each with one-eighth the bandwidth of serial channels.

Each mode has advantages and disadvantages.

The video bit streams are produced in a parallel format, and the cost and complexity of serialization equipment are avoided with parallel transmission. Thus parallel systems are cheaper and simpler for local signal distribution. On the other hand, multiconductor cable is more expensive and bulky, and parallel transmission is more costly for long distance transmission. Matching the delay times of parallel paths may be difficult in parallel systems, and there is the possibility of crosstalk between paths. Matrix switching between multiple inputs and outputs is difficult. Signal processing equipment is more costly and complex for the parallel mode. As a result, serial transmission is widely used where distances are great or for complex signal processing.

In the absence of bandwidth compression, the bandwidth requirements of either serial or parallel transmission, 54 MHz or higher for 8-bit composite or luminance signals, are a serious handicap to the use of the digital mode for long distance transmission. The bandwidth requirement for 8-bit signals does not fall neatly into one of the steps in the standard 1.54-Mbits/s T-1 communications hierarchy, which has levels at 44.7 and 274.2 Mbits/s.

Metallic coaxial cable is impractical for transmission of uncompressed video signals except for very short circuit lengths. The normal bandwidth of C-band satellite and microwave channels is 36 MHz (see Chaps. 11 and 12).

Fiber optics is the best prospect for increased bandwidth. Fiber optic circuits usually employ digital transmission and have circuit capacities upwards of 200 Mbits/s. Bandwidth compression (see Sec. 4.18) is an equally promising solution to the bandwidth problem.

Although bandwidth receives the major emphasis in digital system design, *envelope delay* distortion must be kept within reasonable limits to avoid bit distortion. On the other hand, digital signals are considerably more tolerant of noise than analog signals, and linearity is not a factor.

4.15   Error Correction

A major advantage of the digital mode is its ability, within limits, to recognize and correct bit errors that are caused by noise in the system.

If the number of errors exceeds these limits, an additional technique, *error concealment*, can be applied that reduces the visible effect of the errors. This technique is particularly useful in video recorders.

4.15.1 Visual effect of bit errors

The visual effect of an error depends greatly on the significance of the bit. In an 8-bit system, an error in the least significant bit will be 1 part in 256, with a barely noticeable effect. An error in the most significant bit, on the other hand, will cause a major disturbance in the signal and produce either a white or a black dot. This range in sensitivities is considered in some error correction systems.

4.15.2 Error correction system functions

Figure 4.10 is a functional diagram showing the essential elements of error correction systems.

The redundancy code is a bit or word added to the digital bit stream to make it possible to detect bit errors with a high degree of probability.

The retransmission request function is effective only in systems that have *foreword error correction*. Unlike in computer systems that do not operate in real time, this function is required for video signals if retransmission of a bit that is suspected of being in error is desired.

Since bits have only two states, the correction of an error is straightforward once it has been detected.

Error concealment, applicable primarily to digital recorders, is a more complex process. It is used when error correction is ineffective, e.g., when a large number of bits are in error. Its principle is to reconstruct the missing or erroneous bits by interpolation of adjacent bits.

4.15.3 Encoding and error detection

A variety of redundancy codes have been developed to make it possible to identify bit errors with a high degree of certainty and probability. The simplest code is a single 1 or 0 bit, known as a parity bit, added to

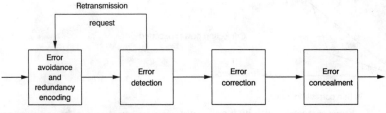

Figure 4.10 Error correction system.

the end of each word as required to make the sum of the bits an even number. This code is illustrated for 2-bit words.

Signal words	Parity bit	Sum
0 0	0	0
0 1	1	2
1 0	1	2
1 1	0	2

If the sum of the bits in a word plus its parity bit after transmission or recording is an even number (0 is considered to be an even number), it is probable that none of the bits is in error, although there is a possibility that more than one are incorrect.

This simple parity check is effective in identifying a single error, but the bit error rate (ber) must be rather low for it to be effective. For larger bit error rates such as are apt to occur in video recording, more complex codes are required.

Another problem with this procedure is that, although it indicates the probability of an error, it does not identify the offending bit so that it can be corrected. This problem can be solved by applying the parity principle to columns as well as rows in the bit matrix. This is known as a *crossword* or *product code*.

More complex and sophisticated codes have been developed that can operate with higher bit error rates and in two dimensions. Two of the most commonly used are the *Hamming code*[1] and the *Reed-Solomon code*.[2] Descriptions of these codes are to be found in the references.

4.15.4 Effectiveness of error correction

Figure 4.11 illustrates the performance of an error correction system as the bit error rate is increased. In a system with no error correction,

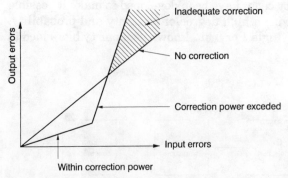

Figure 4.11   Error correction with high error rate.

the output bit error rate increases with the input (assuming that no additional errors are introduced). Error correction is effective for input error rates up to a critical value. Above this level, not only is error correction ineffective, but the output ber is actually worse than without correction. If the input ber exceeds this critical level, the error correction function should be disabled.

4.16   Digital-to-Analog Conversion

Digital video signals must ultimately be reconverted to an analog format to be useful. Figure 4.12 illustrates the principle of one type of digital-to-analog converter. A circuit is clocked so that each bit in a word, from the LSB to the MSB, controls a current flow that is proportional to its position in the word. The summation of these currents is proportional to the quantized level in the analog signal. The analog signal is then recovered by passing the output through a low-pass filter that removes the sampling frequency components.

An obvious problem with this system is the necessity of controlling the current in the MSB path with a high degree of accuracy. With an 8-bit code, the error in the MSB path must be substantially less than 1 part in 256 or it will overwhelm the effect of the LSB.

An alternative conversion system causes a constant current to flow into an integrator for a time interval that is determined by the position of the bit in the word.

The output of these conversion systems is a reconstruction of the pulse train that was created by the sampling process in the A-to-D converter. It is necessary to pass it through a low-pass filter having a cutoff at one-half the pulse rate in order to recover the original analog signal.

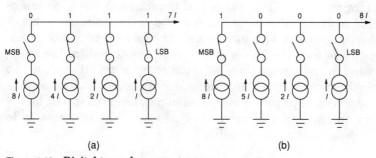

(a)   (b)

Figure 4.12 Digital-to-analog converter.

4.17 Video Bandwidth Compression

For the past several years, the television technical community has been engaged in two major development programs—HDTV and video compression—both directed toward more efficient use of spectrum space but with different objectives. The major purpose of HDTV development is to improve picture quality with a minimum increase in bandwidth requirements. The major purpose of bandwidth reduction programs, sometimes known as video compression, is to reduce bandwidth requirements with a minimum loss of picture quality. In some respects these objectives are complementary, but in others they are in conflict and will require tradeoffs. The industry has yet to reach agreement on the technology that offers the best solution, but the results to date are encouraging. A successful solution, which at reasonable cost simultaneously improves picture quality and reduces bandwidth requirements, will revolutionize the industry.

At the time of this writing (1992), all of the television industry's major laboratories and many smaller ones are carrying out intensive development work in video compression systems. Because of the rapidity and volatility of the developments, the complexity of the technology, and the uncertainty of the ultimate outcome—the specific system or systems that will be standardized by the industry—this chapter describes only the basics of bandwidth reduction systems.

Communications theory[3,4] shows that under the conditions encountered in practical television systems, the rate at which information can be transmitted in a communications channel is proportional to its bandwidth. Bandwidth reduction, therefore, inevitably results in a corresponding reduction in the rate of information transfer and in at least a small loss of image quality.

If television signals were random current or voltage variations, bandwidth reduction would not be possible without a major loss of picture quality. Fortunately, this is not the case. Television signals are highly structured, and they are repetitive in time and correlated in space. Further, the rate of information transfer varies widely from frame to frame and in different areas of the frame. Areas of an image with rapid motion and a large amount of detail require more bandwidth than quiet areas with little detail. The role of bandwidth reduction systems is to make use of these properties to reduce the bandwidth with a minimum loss of picture quality. It should be emphasized, however, that any bandwidth reduction system will result in *some* loss in picture quality—examples are artifacts or loss of resolution on the edges of moving objects—and the objective is to make the defects virtually invisible.

4.18 Digital Bandwidth Compression Technologies

The digital format is uniquely capable of performing the signal processing that is necessary to accomplish the objective of achieving bandwidth reduction with a minimum loss of picture quality. Since the signal consists of a number sequence, it can be transformed continuously to a band-reduced format by a series of algorithms, which are extremely powerful mathematical tools.

4.18.1 Algorithms

As with other complex digital processing functions, the *algorithm* is the basic tool for specifying the mathematical operations that are to be performed for bandwidth reduction. The role of the algorithm in a bandwidth reduction system is shown in Fig. 4.13. It combines the mathematical operations for one or more bandwidth reduction techniques and applies them continuously to the incoming bit stream.

4.18.2 Fixed and adaptive techniques

Figure 4.13[5] is based on five common techniques for video compression:

- Truncation
- Subsampling
- Prediction
- Transformation
- Statistical

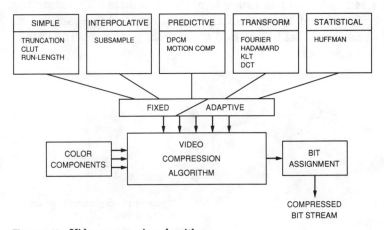

Figure 4.13   Video compression algorithms.

This is by no means a complete list of techniques, but it is illustrative of the range of possibilities.

The techniques have two modes, fixed and adaptive. In the fixed mode, their application is independent of image content. In the adaptive mode, their application depends on the content—for example, they may operate differently for images with a great deal of detail or motion.

The resulting algorithm processes the incoming digital signals continuously, and the output is a bandwidth-compressed bit stream.

4.18.3 Truncation

Truncation is a brute force technique that reduces the bit rate by dropping one or more of the least significant bits (LSBs). In the fixed mode, truncation has the obvious problem that it lacks the ability to distinguish between the brightness levels represented by the LSBs, and the variations in image brightness will be discontinuous. This can be avoided by an adaptive transform that drops the LSBs only to the extent that the amount of picture detail permits.

The bit stream with variable word length is fed into a buffer, after which the missing LSBs are restored. A potential problem with this system is that the buffer can become overloaded if the picture contains an unusual amount of motion or detail.

4.18.4 Subsampling and interpolation

Subsampling and interpolation is a widely used bandwidth reduction technique; it can be used with samples that have not been digitized. Figure 4.14 illustrates one example of the technique.[6] The sampling frequency is reduced by one-half (subsampling), and the samples on adjacent lines are offset by one-half the sampling interval. The values

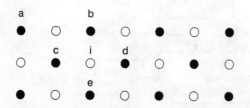

Figure 4.14 Subsampling and interpolation.

● Sampled levels

○ Interpolated levels

of the samples that are skipped as the result of subsampling are interpolated from the values of adjacent sampled elements, both horizontal and vertical. Two possible interpolation equations for the example in Fig. 4.14 are

$$i = \frac{b + c + d + e}{4} \tag{4.3}$$

or
$$i = \frac{c + d}{2} + \frac{b - e}{2} \tag{4.4}$$

Equation (4.3) is the arithmetic average of all the adjacent sample values. This is sometimes called A interpolation. Equation (4.4) is the average value of the adjacent points on the same line plus the *change* in value of points on an adjacent line—a function of the high-frequency components in the signal. The choice between these formulas is largely subjective.

4.18.5 Prediction

The prediction technique is based on the fact that the difference in signal levels between adjacent pixels is usually small. If this is the case, the difference can be represented by a smaller number of bits per word, say 4 instead of 8. This technique is called *differential pulse code modulation (DPCM)*.

A DPCM system is overloaded when there is an occasional large difference between adjacent pixels, as at a very sharp edge. This is called *slope overload* and results in a smeared edge.

DPCM also has the problem that the amplitude of each pixel signal is the sum of the amplitude of the adjacent pixel signal and the difference. If an error creeps into one of the pixels, it will be repeated until the level is reset.

As the result of these problems, DPCM is not widely used by itself, but it is a useful technique when combined with others.

4.18.6 Transformation

Video compression can be accomplished by a *transformation* of the values of a group of pixels into another set, which can be transmitted with less data. After transmission, an *inverse transform* is performed that recovers the original values.

Transforms have been the subject of extended investigation, the objective being to have them be effective and operate speedily in real time. As a simple example, assume the values *A*, *B*, *C*, and *D* are transformed to the values *W*, *X*, *Y*, *Z* by the equations

$$W = A \qquad Y = C - A$$
$$X = B - A \qquad Z = D - A$$

W, X, Y, and Z are transmitted. Since X, Y, and Z are value differences, they are usually smaller than the untransformed originals B, C, and D and do not require as many bits per word.

After transmission, A, B, C, and D are recovered by a *reverse transform:*

$$A = W \qquad C = W + Y$$
$$B = W + X \qquad D = W + Z$$

Adaptive transformation is another technique. The amount of information that must be transmitted for a particular portion of the image is proportional to the fineness of detail in that portion. A portion of the picture with little detail can be transmitted with very few bits, and this provides extra time for transmitting portions with high detail. A buffer is used to restore the original spatio-temporal relationship. If the entire picture has high detail, the buffer may become overloaded and the rate of information transfer is reduced by increasing the quantum intervals. This, of course, reduces the image quality by introducing artifacts.

Transforms have been the subject of extended mathematical analysis, and many are in use. Some of the better known ones are the *Hadamard* transform, the *Fourier* transform (not to be confused with the Fourier transform that converts from the frequency domain to the time domain), and the *discrete cosine function.*

4.18.7 Statistical techniques

Statistical techniques can be used for bandwidth reduction, making use of the fact that some image values are less frequently encountered than others. Shorter words are used to define the more frequently encountered values, with a reduction in the total number of bits. This technique requires the transmission of a code such as the *Huffman code* so that the length of each byte can be identified at the receiving end.

4.18.8 Cascaded compression

To achieve a desired degree of compression, it may be necessary to combine more than one technique and to compress the signal in a number of successive steps. For example, a compression of 128 to 1 could be achieved by 4:1, 4:1, and 2:1 compressions performed successively.

References

1. R. W. Hamming, "Error Detecting and Error Detecting Codes," *Bell Syst. Tech J.*, 26:147–160, 1950.
2. I. S. Reed and G. Solomon, "Polynomial Codes Over Certain Finite Fields,"*J. Soc. Ind. Appl. Math*, 8:300–304, 1960.
3. R. V. L. Hartley, '"Transmission of Information," *Bell Syst. Tech. J.*, 7, July 1928.
4. C. E. Shannon, "A Mathematical Theory of Communication," *Bell Syst. Tech. J.*, 27, July–October 1948.
5. Arch Luther, *Digital Video in a PC Environment*, McGraw-Hill, New York, 1991.
6. R. Kishimoto, N. Sakuri, and A. Ishikura, "Bit-Rate Reduction in the Transmission of High-Definition Signals, *Jour. SMPTE*, 96(2), February 1987.

Television Cameras

5.1 Introduction

The camera is the most fundamental element of television systems, and its most fundamental component is the *imager,* the device that converts the optical image into an electrical signal. Cameras are designed around the characteristics of their imagers, and a description of cameras must begin with imager technology. This technology has changed enormously during the past 45 years, and its development has included three breakthroughs, one of which—the shift from storage tubes to solid-state devices—is still in progress.

5.1.1 A brief history of imagers

The first imagers were *scanners.* They formed a raster on a live scene or film frame by scanning it with a bright spot of light. A phototube picked up the reflected or transmitted light from the raster and generated the electrical signal. The Nipkow disk, the earliest imager, was a mechanical scanner. It was succeeded by the Farnsworth image dissector, an electronic scanner that showed brief promise. *Flying spot scanners* produce a raster on a kinescope and focus it on a frame of film.

All scanners suffer from inherently low sensitivity, which makes them impractical for live pickup.

The first breakthrough in imager technology came with the introduction of *photoemissive storage tubes,* in which incident light from an image of the scene causes electrons to be emitted from a photosensitive surface in a pattern that corresponds to the brightness of the image. These tubes have sufficient sensitivity for live scenes because they store the light energy in an image of the scene for an entire frame before it is removed by the scanning process.

The *iconoscope,* developed by V. K. Zworykin of RCA prior to World War II, was the first practical photoemissive storage tube, and it proved the feasibility of all-electronic television. But it produced a noisy picture, it was difficult to operate (the term "shading" resulted from one of its operational problems), and its sensitivity was marginal for live pickup.

The iconoscope was succeeded for live pickup by the *image orthicon*, a product of World War II research. It had greater sensitivity, and iconoscopes were used only for film pickup in the postwar era. The image orthicon had good sensitivity, but it was also difficult to operate (although easier than the iconoscope), and its signal-to-noise ratio was marginal.

The defects of the iconoscope and image orthicon led to the second breakthrough, the replacement of photoemissive tubes with *photoconductive* types. Photoconductive tubes make use of changes in the apparent electrical resistance of a *photoconductor* when exposed to light. They are smaller, cheaper, easier to operate, and more noise-free than the iconoscope or image orthicon. The first commercial photoconductive tube was the *vidicon*, which was introduced for film pickup in the early 1950s. Its sensitivity was marginal for live pickup, but its use for film continued for more than 30 years.

Ten years later, a new photoconductive tube, the *Plumbicon*, was introduced. It had sufficient sensitivity for live pickup as well as the other desirable features of photoemissive tubes, and by the early 1970s it had replaced the image orthicon. Another photoconductive tube, the *Saticon*, was introduced in 1974.

As of this writing (1991), the third breakthrough is under way, as photoconductive tubes are giving way to solid-state devices called *charge-coupled devices* or CCDs. This is one more step in the replacement of vacuum tubes by solid-state devices in the electronics industry. Storage tube imagers still have a place, however, and this chapter describes both technologies.

Table 5.1 summarizes the introduction dates and technologies of the principal types of storage tubes and CCDs. The introduction dates are not precise, but indicate the approximate time period when the imagers became standard commercial items.

TABLE 5.1   Principal Television Imagers

Name	Introduced	Type	Application
Iconoscope	1939	Photoemissive	Film/Live
Image orthicon	1946	Photoemissive	Live
Vidicon	1952	Photoconductive	Film/Live
Plumbicon	1963	Photoconductive	Live/Film
Saticon	1974	Photoconductive	Live/Film
CCD	1980	Solid-state	Live/Film

5.1.2   Camera categories

There are two major television camera categories, live and film.

Live cameras generate video signals from the optical images of indoor and outdoor scenes. They are classified as studio, outside broadcast (OB), and portable. Studio cameras are designed to provide optimum performance in a controlled indoor environment, usually at the expense of size, weight, and portability. The performance of OB cameras often approaches or equals that of studio cameras, but the packaging of their components is designed so that they are *transportable,* if not *portable.* Cameras designed for electronic news gathering (ENG) or satellite news gathering (SNG) provide portability, even to the extent of being hand-held. In some cases, a portable camera is combined with a video recorder; the combination is called a *camcorder.*

Film cameras generate television signals from film or slide images. The utilization of film as a source of video programming is diminishing, since most programs are now produced and recorded on tape. Existing film libraries are usually transferred to tape for distribution. And the advent of highly portable TV cameras (see Sec. 5.13) has caused a shift from film to videotape for on-the-spot news gathering. There remain, however, limited but vital applications for TV film cameras, especially film-to-tape transfers.

5.1.3   Live camera configurations

The configuration of apparatus required to produce a video signal from an optical image is known as a *camera chain.* For studio cameras, it consists of the *camera head* together with the rack- or console-mounted units that perform the control, signal processing, composite signal multiplexing, and monitoring functions. ENG cameras may be self-contained, or they may have a portable control unit that is used for setup and optionally for operation. Figure 5.1 is the functional diagram of a studio color camera chains.

The basic functions of ENG and studio cameras are similar, but the ENG camera must have the ability to control operating parameters such as gain and sensitivity automatically. The camera is set up initially with the use of external monitoring equipment, but its stability and automatic features permit it to operate for considerable periods without manual change of the setup adjustments.

5.1.4   Film camera configurations

Film camera systems must cope with the incompatibility of the standard 24-frame-per-second film rate and 25 frames per second for most PAL systems or approximately 30 per second for NTSC. The difference

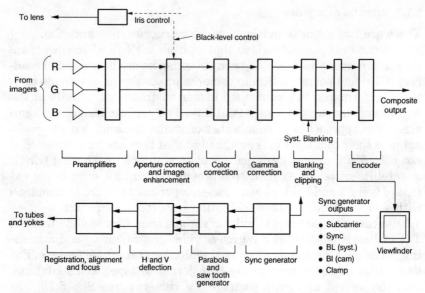

Figure 5.1 Studio camera chain. (*From K. Blair Benson,* Television Engineering Handbook, *McGraw-Hill, New York, 1986, Fig. 5.1.*)

is usually ignored for PAL, and the film is simply run at the higher rate. The resulting distortion is barely visible for video and noticeable only to the rare individuals with perfect pitch for audio.

For NTSC systems, the projector/scanner combination must have the facility of converting from 24 to 30 frames, and a number of complex electronic and mechanical systems have been developed to accomplish the conversion. These are now being replaced by digital frame store devices similar to those used for standards conversion (see Sec. 4.5.2). The signal is generated at 24 frames per second, converted to digital form, stored in a digital memory, and read out at 25 or 30 frames per second, thus making even the small correction required for PAL systems.

5.2   Photoconductive Storage Tubes

The *vidicon* was the first of a family of commercial photoconductive tubes that eventually included *Plumbicons, Saticons, Newvicons, Chalnicons,* and *silicon diode* tubes. The term *vidicon* is sometimes used generically to indicate any photoconductive tube.

The vidicon utilizes the photoconductive properties of antimony trisulfide, and it is an excellent tube for film pickup. It suffers from severe lag or smear, however, when its operating conditions are adjusted for high sensitivity, and its use for live pickup is primarily limited to industrial applications.

The Plumbicon is a *heterojunction* photoconductive tube that uses a lead oxide semiconductor as the photosensitive material. It greatly reduced the vidicon's lag problem and is now widely used for live pickups.

The Saticon is a photoconductive tube that uses a selenium alloy semiconductor as the photoconductive material. It was introduced later than the Plumbicon and enjoys certain advantages, but both continue to be used. Whereas the response of the Plumbicon to red light is higher after special doping of the semiconductor, the response of Saticons is higher in the blue end of the spectrum.

Vidicons, Plumbicons, and Saticons are sometimes sold under other trade names.

In addition to these three principal types of photoconductive tubes, there are others that are used for special purposes. They include the *Newvicon,* which uses zinc selenide as the photoconductor, the *Chalnicon,* which uses cadmium selenide, and silicon diode tubes. These tubes have greater sensitivity, particularly in the red and infrared portions of the spectrum, and are used mainly for surveillance cameras.

5.2.1 Photoconductive tube construction

The construction of photoconductive storage tubes is shown in Fig. 5.2.

The faceplate, mounted on the front of the tube, is a glass disk with its inner surface coated with a thin layer or layers of photoconductive material. A ring of conductive material is sealed to its circumference, forming a structure known as the *target.* The target is sealed to the front end of the tube, and a connection to the conductive ring is the *signal electrode.*

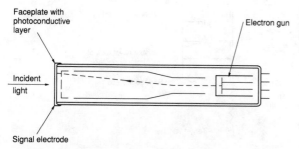

Figure 5.2 Photoconductive storage tubes. (*From K. Blair Benson,* Television Engineering Handbook, *McGraw-Hill, New York, 1986, Fig. 20.18.*)

The photoconductive surface for vidicons is a layer of antimony trisulfide (Sb_2S_3) on a thin transparent substrate. For Plumbicons it is a layer of lead oxide (PbO) which is treated to have I-type or *heterojunction* semiconductivity, sandwiched between a layer of tin or indium oxide with P-type semiconductivity and a second layer of lead oxide treated to have N-type semiconductivity. For Saticons it is a layer of P-type selenium arsenic sandwiched between a layer of antimony trisulfide on the scanned side and successive layers of selenium tellurium and N-type selenium arsenic on the image side.

An *electron gun* at the rear of the tube generates a very narrow beam of low-velocity electrons that is made to scan the target in a raster. The beam is focused, either electrically by means of electrodes in the electron gun or magnetically by a focus coil with an axis concentric with the tube. Similarly, the deflection can be either electromagnetic, by deflection coils having their axes at right angles to the beam, or electrostatic, by deflection plates within the tube.

5.2.2 Photoconductive tube operation

The basic operation of all photoconductive tubes is similar, although the mechanism is quite different. An optical image is formed on the photosensitve surface of the target. The target is scanned by a low-velocity electron beam, and the image is converted to an electrical signal. The principle of this process for the vidicon is shown in Fig. 5.3.

Each pixel on the target acts as a capacitor with a leaky dielectric, the signal electrode forming one plate and the inner surface of the

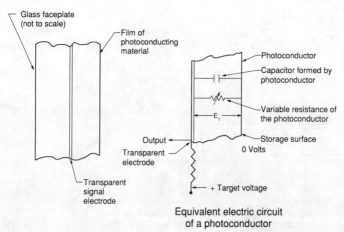

Figure 5.3 Photoconductor operation. (*From K. Blair Benson,* Television Engineering Handbook, *McGraw-Hill, New York, 1986, Figs. 11.10 and 11.11.*)

photoconductor the other. A positive voltage, the target voltage, is applied to the signal electrode through the load resistor. When the pixel is not illuminated and its resistance is high, its inner surface assumes the cathode potential as the result of scanning, and the pixel capacitance is charged to the target voltage. When the pixel is illuminated, the resistance of the photoconductor at the pixel location is reduced, and its inner surface becomes more positive after being scanned as it rises toward the target voltage. When the pixel is scanned again, electrons are absorbed from the beam to restore the inner pixel surface to the cathode potential by recharging the pixel capacitance through the load resistor. The magnitude of the current through the load resistor and the signal voltage are determined by the amount of charge that leaked through the pixel capacitance during the frame interval between scans. This, in turn, is determined by the illumination of the pixel, and the signal voltage is determined by the illumination of the target.

The resistance of the photoconductor changes with illumination because the energy of the light photons is sufficient to liberate electrons from individual atoms, forming free electrons and positive holes. The free electrons migrate to the positively charged target electrode, and the holes migrate to the scanned surface.

The individual pixels on the photoconductor are not insulated from one another—the photoconductor surface is continuous, and pixels have only a mathematical existence—but the resistance of the photoconductor is so high that lateral current leakage is small. It is large enough, however, to require a careful choice of the photoconductor's thickness. The thinner the photoconductor, the lower the leakage and the greater the definition. On the other hand, the thinner the photoconductor, the greater its capacitance and the greater its lag (Sec. 5.4.7).

Although the conductivity of the photoconductor is low in the absence of illumination, it is not zero, and there is some *dark current*. This is an undesirable effect that is much lower in Plumbicons and Saticons.

In a broad sense, vidicon (Sb_2S_3), Plumbicon (PbO), and Saticon photoconductors are similar in that their apparent resistance is determined by their illumination. The mechanisms by which they operate, however, are different. Vidicon photoconductors operate by changes in ohmic resistance, while Plumbicons and Saticons employ semiconductive materials with back-biased junctions so that there is virtually no dark current, a serious defect of vidicons.

5.3 Charge-Coupled Device Imagers (CCDs)[1]

The use of solid-state photosensors in television began on an experimental basis in the 1960s and was accelerated by the invention of *charge-coupled device* (CCD) imagers in 1969. CCDs have had important uses in nontelevision applications, including ultra-sensitive detectors in astronomy and military surveillance.

A CCD imager consists of rows and columns of pixel-sized solid-state photosensitive elements mounted in a rectangular array on a silicon substrate. They accumulate and store electric charges that are proportional to their illumination from an image focused on the array. Then, either directly or through intermediate *shift registers,* the charges are transferred from element to element through the substrate to the device output in the proper time sequence to generate the video output signal. The transfer function is similar to that of a scanning beam in a photoconductive tube—the conversion of the electric charge pattern on a spatial array of photosensors into a television signal.

Strictly speaking, "CCD" refers to the component that performs the transfer function, but it is commonly used to describe the entire imager.

CCDs moved from the laboratory to commercial cameras in the 1970s, initially for portable and field applications where their low lag, low power drain, ease of operation, and small size were particularly important. Currently available CCD imagers have important advantages over storage tubes, both inherent and potential, which make them competitive for studio cameras as well.

A direct comparison of the advantages and disadvantages of storage tubes and CCDs as imagers is necessarily imprecise because of variations in models and operating conditions. Table 5.2 provides a qualitative comparison of the performance of the two types of imagers of current design. More quantitative comparisons are included in subsequent sections. The advantages of CCDs can be expected to increase in the future because of continuing development effort. Although the performance of tubes has also improved and is a moving target, as of this writing (1991) it appears that CCDs will eventually supplant tubes as imagers for all types of cameras.

5.3.1 CCD photosensor arrays

The sensing function of CCD imagers is performed by an array of pixel-sized photodiodes. Each photodiode develops a charge that is proportional to the illumination of the image at that point. This function of the photodiode is combined with the storage and charge-transfer function of the CCD.

The resolution of a CCD camera is determined by the number of photosensitive elements. Since each row of these elements provides the signal for one scanning line, the number of rows should at least equal the number of scanning lines. The number of elements in each column will determine the horizontal definition in accordance with sampling theory and will determine the severity of aliasing (Sec. 4.3.4).

TABLE 5.2   Comparison of Storage Tube and CCD Imagers

Parameter	Comparison
Resolution and definition	Tubes have higher resolution than present CCDs, but aperture response of CCDs at lower line numbers is better
Lag and burn-in	Major advantage for CCDs
Smear	Major problem for CCDs, but solutions exist
Dynamic resolution	Advantage for CCDs because of absence of lag
Signal-to-noise ratio	S/N for tubes depends on preamplifier, while noise in CCDs is generated internally (see Sec. 5.4.4); small advantage for CCDs
Artifacts	Advantage for tubes
Sensitivity	Small advantage for CCDs
Overload blooming	Advantage for tubes
Colorimetry	No major difference with color correction
Sensitivity to vibration	Advantage for CCDs
Registration	Advantage for CCDs
Ease of operation	Advantage for CCDs
Life	Advantage for CCDs
Power consumption	Advantage for CCDs
Size	Advantage for CCDs

A CCD intended for broadcast use typically has 492 rows, the number of active scanning lines in an NTSC transmission, and 510 columns. While this number of elements is adequate for broadcast standards, it is not satisfactory for HDTV, which would require approximately double the number of rows and columns, or four times the number of elements. The obvious solution, the construction of a CCD chip with smaller elements, has two problems. The first is manufacturing, particularly the yield problem with four times the number of elements that can have defects. The second is lower sensitivity, which results because the charge generated by each element is proportional to its area. These problems are being solved, however, and it seems reasonably certain that CCDs will be used for HDTV when it becomes commercially practical.

5.3.2 CCD transfer and readout architectures

The essence of CCD imagers is their method, often called their *architecture,* of converting a spatial charge pattern on an array of photosensors into a time-varying video signal. Three commonly used architectures are *frame transfer* (FT), *interline transfer* (IT), and *frame interline transfer* (IFT).

5.3.2.1 Frame transfer architecture.

Frame transfer architecture is illustrated in Fig. 5.4. The CCD structure is divided into three sections, a sensor array, a storage register array, and a horizontal output register. Charges are allowed to accumulate at each photosensor for a complete frame. At a command from a clock at the end of the frame, the charges in each column of pixels in the sensing area are shifted to a corresponding column in the storage area. This frees the sensors to accumulate charges from the illumination of the next frame. From the storage area, again at the command of a clock, the charges are shifted one line at a time to the output register that generates the video signal.

Frame transfer architecture suffers from an inherent problem of overload of the pixels' storage units in excessively bright sections of the image. This results in vertical smear and blooming.

An effective solution to the smear problem is to provide a mechanical shutter that shields the photosensitive area from illumination during the transfer period, but this has the disadvantage of introducing a mechanical device into an otherwise all-electronic product.

Blooming and smear can be reduced by the use of frame interline transfer (Sec. 5.3.2.3).

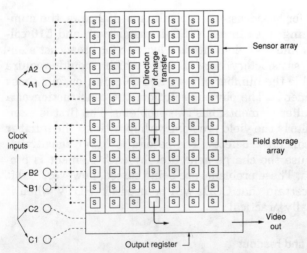

Figure 5.4 Frame transfer architecture. (From Donald G. Fink and Donald Christiansen, Electronic Engineers' Handbook, 3d ed., *McGraw-Hill, New York, 1989, Fig. 20.44a.*)

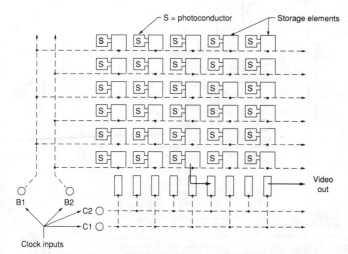

Figure 5.5 Interline transfer architecture. (From Donald G. Fink and Donald Christiansen, Electronic Engineers' Handbook, 3d ed., *McGraw-Hill, New York, 1989, Fig. 20.44b.*)

5.3.2.2 Interline transfer architecture. The operation of interline transfer architecture is illustrated in Fig. 5.5.

Vertical columns of photosensors are connected to alternate columns of CCD transfer registers. After exposure to a single frame and when commanded by a clock, the photo-generated charges in the pixels in each column are transferred along their rows to the neighboring vertical transfer register. This frees the photosensors to accumulate new charges as determined by the illumination of the next frame.

Another clock command causes the charges to be shifted from the vertical register to the horizontal register. From this register, they are read out as the video signal.

In the basic form of this architecture, the vertical register columns must be covered with opaque masks, and only 30 to 50 percent of the photosensor area is photosensitive. More complex variations of this architecture, such as rear-illuminating the chip, have substantially increased the light efficiency.

5.3.2.3 Frame interline transfer architecture. Frame interline architecture was developed to provide the advantages of frame transfer, but with resistance to smearing and blooming from excessive highlights. The unique feature of frame interline transfer architecture is a row of

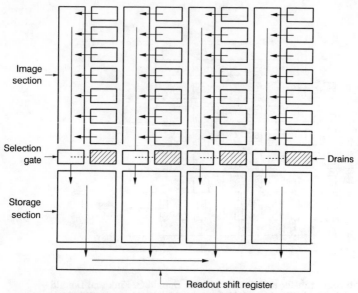

Figure 5.6 Frame interline transfer architecture.

selection gates between the image and storage areas, as shown in Fig. 5.6. The gates are biased so that charges in excess of a predetermined level are drained from the system before being transferred to the storage area.

5.4   Storage Tube and CCD Performance

This section compares the performance of vidicons, Plumbicons, Saticons, and CCDs with respect to aperture response and resolution, signal-to-noise ratio, sensitivity, lag, transfer function (gamma), and spectral response. The performance of individual components under different operating conditions varies considerably, and the comparisons are approximate rather than precise.

5.4.1   Storage tube aperture response and resolution

The aperture response of a standard 1-in vidicon expressed in terms of its modulation transfer function (MTF) and contrast transfer function (CTF) is shown in Fig. 2.4. The MTF at 400 lines is about 33 percent. The MTF of original 1.2-in Plumbicon tubes was comparable.

Subsequently, motivated by competition, the desire for smaller tubes, and the future demands of HDTV systems, there has been a significant increase in the aperture response of all types of photoconductive tubes. The improvements leading to the increased MTF include better electron optics and the use of diode guns that reduce the lag. These improvements have made it possible to obtain satisfactory results for broadcast purposes, even in studio cameras, with smaller tubes, and ⅔-in Plumbicons and Saticons are commonly used. They also can be used to increase the definition of larger storage tubes as required for HDTV service.

5.4.2 CCD aperture response and resolution

Figure 5.7 shows the aperture response of CCDs with two different element widths, one equal to the element spacing and the other one-half the element spacing. The Nyquist limit is the number of sensors per picture height. The narrower element width gives better aperture response, but wider elements result in less aliasing and greater sensitivity.

Since CCDs utilize a sampling process in both dimensions, the ambiguities that are inherent in specifying and measuring vertical resolution in all television systems are present for both vertical and horizontal resolution. The width of the photosensitive elements adds another major variable that does not exist with phototubes. In fact, there is a question as to the significance of resolution as a criterion of CCD performance. Nevertheless, the industry is so accustomed to using resolution as a primary specification of imager performance that it is used for CCDs as well.

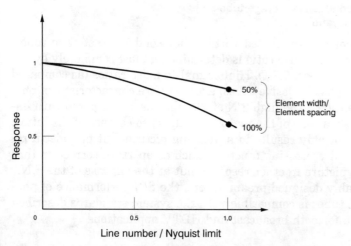

Figure 5.7 Sampling aperture response. Ref. Watkinson, 2.6(b). (*From John Watkinson*, The Art of Digital Video, *Focal Press, London, 1990, Fig. 2.6(c).)*

The resolution of CCDs can be approximated by multiplying the number of elements per picture height by the Kell factor. For the 492 by 510 sensor described above and a Kell factor of 0.7, the vertical resolution would be 344 lines (492 × 0.7) and the horizontal 267 lines (510 × 0.7 × 0.75).

The number of rows of elements should equal the number of active scanning lines—for NTSC approximately 490 and for PAL approximately 580. For equal vertical and horizontal definition, the number of columns should equal the number of active scanning lines times the aspect ratio, or 490 × 1.33 = 655 for NTSC cameras. Most commercial CCD imagers do not yet meet the 655-column standard; 500 is a more typical figure.

The measured horizontal resolution for color cameras can be increased by staggering the horizontal locations on the scanning lines of the photosensitive elements in the R, G, and B imagers. When the outputs are added in the Y-channel matrix, smaller detail can be resolved in the combined output than on any single channel. The effectiveness of this depends, of course, on the use of elements that are relatively narrow in the horizontal direction.

As of this writing (1991), the definition of CCDs is marginal for HDTV applications because of economic and technical limitations on the number of photosensitive elements that can be placed on a single chip. Rapid improvements are being made, however, and it is expected that HDTV cameras employing CCDs will become available at reasonable cost in the not-too-distant future.

5.4.3 Photoconductive storage tube signal-to-noise ratio

Very little noise is generated within photoconductive storage tubes, and their signal-to-noise ratio is determined almost completely by the preamplifier (see Sec. 5.7.1). High signal-to-noise ratios (as compared with those of photoemissive tubes) are therefore characteristic of photoconductive tubes. The high S/N of Plumbicons as compared with image orthicons was one of the reasons for the obsolescence of the latter. A high S/N not only results in a pleasing picture, but provides a reserve for signal processing functions such as aperture correction that improve the picture in other respects, but at the expense of the S/N.

With suitably designed preamplifiers, the S/N performance of photoconductive tubes is compatible with the system standards described in Sec. 2.16.5 for both broadcast and HDTV applications.

5.4.4 CCD signal-to-noise ratio

Unlike in photoconductive tubes, most of the noise from solid-state imagers is generated within the device rather than in the external circuitry. This is indicated by the relative signal currents, which in CCDs are measured in milliamperes, as contrasted with nanoamperes in photoconductive tubes. As a result, the signal-to-noise ratio of modern CCDs, operating with adequate lighting, can meet exacting broadcast standards (Sec. 2.16.5), typically 55 dB (unweighted).

5.4.5 Transfer function (gamma)

Transfer function and gamma as applied to television signals are defined in Sec. 2.15.1.

Plumbicons and Saticons have gammas of unity, while vidicons have a gamma of about 0.65. The lower gamma of the vidicon is an advantage because it is complementary to the high gamma of kinescopes. The transfer characteristic of CCDs is essentially linear.

The dynamic range of imagers is limited on the low end by the signal-to-noise ratio and on the high end by beam current or resolution limitations. As a result of their excellent signal-to-noise ratio, which permits operation down to low light levels, the dynamic range of CCDs is excellent, approaching 1000:1—more than is needed in a practical system. In addition, the black stretch introduced by gamma correction can be introduced without excessive noise.

5.4.6 CCD aliasing

The problem of aliasing in analog systems employing photoconductive tubes was discussed in Sec. 2.3.2. Aliasing is a more severe problem with CCDs because the photosensitive surface inherently performs a sampling function and because there is no optical equivalent of a low-pass, anti-aliasing electrical filter.

Aliasing occurs for stationary scenes when there are spatial frequency components in the image that are above the Nyquist limit. High-frequency components in an electrical signal can be removed by filtering, but the equivalent optical filters, defocusing and diffusion, do not have sharp cutoffs, and significant loss in fine detail results. This is not an especially serious problem for broadcast transmissions (no worse for horizontal detail than that caused by the scanning pattern), but it is potentially serious for HDTV.

Aliasing also results from movement of the image, whether because of movement of the object or panning of the camera. Since photoconductive tubes have no discontinuities in their photosensitive surfaces, there is sufficient smoothing of the motion of edges to avoid aliasing. With the discontinuities in their photosensitive surfaces, CCD cameras sometimes produce images that display aliasing effects—spurious or irregular motions for moving objects.

The problem of aliasing with CCDs makes increasing the number of scanning lines in HDTV systems particularly important.

5.4.7 Lag, smear, and blooming

Lag, smear, and blooming are symptoms of the same general problem: residual or spurious charges that are introduced into the signal.

Lag is characteristic of photoconductive tubes and occurs when the pixel capacitances are not completely recharged by a single scan after illumination is removed from the photosensitive surface. The result is a loss in the definition of images of moving objects and, in extreme cases, smearing and comet tails. It is a serious defect, and it has removed Sb_2S_3 vidicons from contention as imagers for broadcast-quality live color cameras, which must operate with marginal lighting.

The operation of vidicons is a tradeoff between lag and sensitivity. The sensitivity can be increased by increasing the target voltage, but this also increases the lag. If the target voltage is increased sufficiently to operate a live camera, the resulting lag is unsatisfactory for broadcast use except with very bright scene lighting.

The magnitude of the lag can be specified by a plot of the residual signal as a function of the number of fields after light is removed. Alternatively, lag can be specified by a single point on this curve, e.g., the residual signal after the third field. Typical values are 1 to 5 percent.

Lag is not a serious problem with vidicons when they are used in film cameras where adequate light is available, and they are frequently used in that application.

Compared with that of photoconductive tubes, the lag in CCDs is low. One of their first applications in commercial television was in the 1984 World Series, when a CCD camera was focused on rapidly spinning curve balls. The lag was so minor that the stitches on the ball were visible. On the other hand, smear and blooming from excessive highlights are problems that are characteristic of CCDs. Their causes and solutions are described in Sec. 5.3.2.

5.4.8 Sensitivity

There are a number of definitions of the sensitivity of television imagers.

The definition of greatest interest to scientists is their *quantum efficiency,* the percent of incident light quanta that create electron-hole pairs in the photoconductor. For Plumbicons and Saticons, the quantum efficiency is approximately 85 percent at the peak of their spectral response curves.

Three other definitions are used for engineering purposes:

1. The output signal current as a function of faceplate illumination.

2. The speed of the optical system required to produce an output signal of normal amplitude and signal-to-noise ratio with a specified scene brightness. For Plumbicon and Saticon cameras, a typical figure is $f/4$ with 2000 lux illumination and 90 percent reflectance.

3. The minimum illumination required to produce satisfactory (but not optimum) signal quality with the camera lens set at maximum speed and full gain in the video channel. A typical figure for Plumbicons and Saticons is 10 lux with an $f/1.2$ optical system and a video gain of 24 dB.

Recent improvements in CCD technology have increased the sensitivity of CCD images to the point that it is comparable to or even greater than that of Plumbicons and Saticons.

5.4.9 Spectral response

The spectral responses of photoconductive tubes and a CCD of recent design are shown in Fig. 5.8. The response of the photoconductive tubes in this example is lower in the red end of the spectrum because the photons in red light do not have as much energy, and hence they create fewer electron-hole pairs. This effect was particularly severe in early Plumbicons, but their response to red light

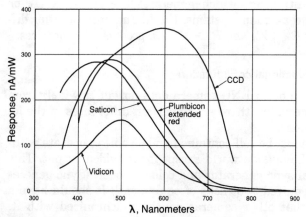

Figure 5.8 Spectral response of television imagers.

has been improved by "doping" the semiconductor, i.e., by deliberately adding impurities.

5.5 Camera Chain Components

5.5.1 Studio camera functional configuration

Figure 5.1 is a functional diagram of a studio camera chain showing its principal components. This diagram is of a camera employing tubes, but the diagram of a CCD camera would be similar except that the sync generator would drive clocks rather than deflection yokes.

The lens, almost invariably a zoom lens, focuses an image of the scene on the R, G, and B imagers, either photoconductive tubes or CCDs, through a beam-splitting prism that separates the light into its three primary components. The outputs of the three imagers pass into preamplifiers, which perform a number of functions such as black level shading in addition to amplification. Following the preamplifiers, the signals are subjected to additional processing functions, including aperture correction and image enhancement (see Sec. 2.10), correction or *masking* by matrixing (see Sec. 3.8.2), and gamma correction. In addition to gamma correction, many cameras have an adjustable knee that limits the output to a fixed preset level to avoid overloading the system.

Finally, the gains of the channels are balanced and sync and blanking are added.

The signal can be monitored and adjusted at all critical points by means of a waveform and picture monitor.

In composite systems, the three processed color signals, R', G', and B', pass to the multiplexer, which generates the NTSC, PAL, or SECAM output. In component systems, the luminance and color difference signals, Y', $R' - Y'$, and $B' - Y'$, are produced by matrices.

5.5.2 ENG camera functional configuration

A functional diagram of an ENG camera does not differ widely from that of a studio camera, but there are important differences in the locations of the functions.

An ENG camera must be self-contained and be able to operate without connection to the outside world except for its video output. This requires an internal sync generator and color encoder. (Sync generators of the 1940s required two racks of equipment!) It should have a *genlock* output so that other cameras can be synchronized with it. CCD cameras, with their lower power requirements, can operate from batteries. Of equal importance, they must be equipped with automatic features so that they can operate over a reasonable range of lighting conditions without adjustment for a reasonable period of time. Manual control from a *camera control unit* (CCU) should be optional.

5.6 Optical Components

5.6.1 Lenses

The technology of television camera lenses has advanced almost as rapidly as electronic technology during the past 30 years. Thirty years ago, the quality of continuously variable focal length *zoom lenses* was not adequate for color cameras, and early color cameras used fixed focal length lenses mounted in a turret. The focal length could be varied only in fixed increments and by interrupting the signal. The widespread use of zoom lenses began in the early 1960s, and they have greatly enhanced the versatility of cameras for program production. Their quality has improved steadily, and their performance now approaches that of high-quality fixed focal length lenses.

Zoom lenses are available for both studio and ENG cameras, with speeds ranging from $f/2.2$ to $f/1.2$ and focal length ratios ranging from 12:1 to 30:1. Lenses for ENG cameras are much smaller, in part because of lower speeds and in part because of the use of smaller imagers, e.g., $\frac{2}{3}$-in Plumbicons.

In cameras employing storage tubes, the effect of geometric distortion can be minimized by the use of *lens error correction*. This is deliberate distortion of the scanning linearity to compensate for lens aberrations. This feature is particularly useful for color cameras because of the need for precise registration. The aberrations of the lens in use are stored in a digital memory, and the linearity of the horizontal and vertical sweep circuits is distorted in the opposite direction to compensate.

A *diascope* is a useful feature of many lenses. It is a built-in test pattern that allows the camera to generate test signals for registration and other setup adjustments without the use of an external test pattern. In modern cameras, it enables many of these adjustments to be made automatically.

The zoom function is controlled electrically, and the focal length can be adjusted either by the camera operator or remotely.

5.6.2 Optical beam splitters

Early color cameras employed image orthicon imagers (see Sec. 5.1.1) that required a complex array of lenses and mirrors to split the light into its three primary components. The advent of the smaller Plumbicon tube made it possible to employ prisms rather than mirrors for light splitting; see Fig. 5.9. Prism light splitters simplify the opti-

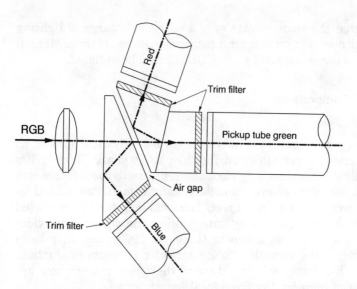

Figure 5.9 Optical beam splitter. (*From K. Blair Benson,* Television Engineering Handbook, *McGraw-Hill, New York, 1986, Fig. 14.41.*)

cal system, they occupy less space, and they are about one f stop faster than mirror systems. Mirror systems also had advantages, however. They made it possible to produce a black surround around the image for a reference black, and the tubes were mounted in a parallel orientation, thus minimizing minor registration problems resulting from the earth's magnetic field. The advantages of prism systems are so great, however, that they are almost universally used.

The operation of the beam splitter can be seen by an examination of Fig. 5.9. The key technology is the *dichroic* surface, which reflects light in one region of the spectrum and transmits the rest. The dichroic surface is coated with a monomolecular film that is one-quarter wavelength long at the center of the reflected spectral band. The phase of reflected light in this portion of the spectrum is shifted 90° in each of its two traversals of the surface and 180° upon reflection, thus emerging in phase.

The spectral passbands of the reflected light are broad, and it is necessary to employ trim filters to sharpen their edges. With trim filters, fairly close conformance to the standard primaries can be achieved by purely optical means, although additional correction (masking) by matrix circuits is usually employed.

As noted earlier, many lens assemblies are equipped with neutral-density filters to reduce the variations in f stop of the lens, and hence depth of focus, with widely varying scene brightnesses.

5.6.3 Neutral-density and color filters

Some cameras are equipped with *neutral-density* and color filters to compensate for wide variations in lighting conditions without excessive adjustment of operating controls.

Neutral-density filters are rated by the amount of light that they transmit, typically ¼ and ¹⁄₁₆. They are used under conditions of intense sunlight.

The spectral response of color filters is specified by their *color temperature,* a term borrowed from photography. As an object is heated, it will begin to glow, first with reddish light and then becoming more bluish as its temperature is increased. The spectral response of a color filter can be specified by the temperature in kelvins (degrees Celsius plus 273°) at which the spectral content of a glowing body most closely matches it.

Typically, the color temperature of incandescent light is 3200 K, while daylight is 5600. The human eye, with its marvelous power of adaptation, automatically compensates for this over a considerable range of temperatures, and the difference between natural and artificial illumination is not noticed when viewing a scene. This power is lost when viewing a television image, and pictures of a scene illuminated by incandescent light from a camera adjusted for daylight will appear reddish. The purpose of the color filters is to compensate for differences in the color temperature of scene illumination.

The color temperatures of a typical complement of color filters are 3200 and 5600. These permit the camera to switch from daylight to artificial illumination without major readjustment of the electronic controls.

5.7 Camera Electronics

5.7.1 Preamplifier

The requirements for storage tube and CCD preamplifiers are quite different because of the different levels of their output currents. The low signal current from a photoconductive tube requires the use of a load impedance with a high resistive component. This results in a falloff in the high-frequency response of the input stage because of the combined shunt capacitance of the photoconductive tube output, the preamplifier input, and their associated circuitry. The falloff must be equalized with a high-peaker circuit, which increases the amplitude of the high-frequency noise components relative to the signal amplitude and reduces the signal-to-noise ratio.[2] The shunt capacitance is minimized by the use of a field-effect transistor—which has a low input capacitance—located within the yoke assembly for the input stage.

The noise spectrum is triangular, i.e., the amplitude of the noise components increases linearly with frequency because of the high-peaker circuit. As a result, the noise weighting factor is large—in excess of 10 dB.

Most of the noise in CCD imagers originates in the imaging device, and no special precautions are required in the preamplifier.

Parabolic and sawtooth waveforms, synchronized with horizontal and vertical scanning, can be added at the preamplifier to compensate for shading errors in the imagers and spurious internal reflections in the optical system.

Camera blanking, i.e., the removal of the signal during the blanking interval, can also be added at the preamplifier. This is augmented by the addition of *system blanking,* which determines the *setup,* or system black level, later in the chain.

5.7.2 Aperture correction and image enhancement

The gain-frequency characteristics of aperture correction and image enhancement circuits are described in Sec. 2.10. The gain of higher-frequency signal components is increased sufficiently to make the aperture response of the camera flat to about 400 lines. Additional correction, known as image enhancement, can be used in the 200- to 300-line region to increase the sharpness of the picture further, but it must be applied with care to avoid unnatural overshoots.

In one refinement of the aperture correction process, it is applied only in the midrange of the gray scale. This avoids degradation of the S/N in the dark areas of the picture, where noise is most visible, and also reduces a tendency to make skin tones rough.

It is important that the aperture correction gain enhancement be achieved without phase shift to avoid enhancing overshoots. The *transversal filter*[3] is one means of achieving this.

Contours-out-of-green is another technique for increasing picture sharpness. Pulses are generated from the green signal at each edge in the image by separating its high-frequency components. The green channel is used to generate these pulses because it has the best signal-to-noise level. The "white" contours—changes from dark to light—are added at the aperture correction circuit, while the "black" contours are added after gamma correction.

5.7.3 Color correction (masking)

Electronic color correction is employed for two purposes: to correct residual errors in the colorimetry of the optical system and to introduce deliberate distortions of the colorimetry for aesthetic reasons. Electronic correction is accomplished by matrixing, and is sometimes called *masking*.

5.7.4 Gamma correction and adjustable knee

Gamma correction, about 0.45, is introduced in each of the three color channels to compensate for the high gamma of the receiver picture tube. This stretches the signal in the black area where noise is very visible, and the uncorrected signal must have a good signal-to-noise ratio.

Additional gamma correction can be applied to the luminance, Y, channel of the encoder. This has the advantage that it does not upset the color balance of the signal.

To handle extremely large variations in scene brightness, some cameras provide an adjustable "knee" in the transfer characteristic. The characteristic follows the normal gamma curve up to an adjustable breakpoint, above which increases in illumination produce only a minor output in signal level.

5.7.5 Sync, blanking, and clipping

The sync generator can be either external, as in a studio installation where cameras, recorders, and editing and processing equipment are synchronized by a common source, or internal, as in a self-contained field camera. The *genlock*, a device that derives its timing pulses and subcarrier from an external signal, is a third possibility.

The system blanking and clipping function establishes the final system black-level setup. This is 7.5 percent for NTSC, the original monochrome standard, and 0 for PAL.

5.7.6 Encoder

Encoding, the conversion of the three primary color signals into composite NTSC, PAL, or SECAM signals, is the final function of the camera chain. The characteristics of the encoded signals were described in Chap. 3.

Figure 5.10 is a block diagram of an NTSC encoder. The gamma-corrected primary color signals, E'_R, E'_G, and E'_B, are passed through matrices to form the E'_Y, E'_I, and E'_Q signals.

Sync is added to the E'_Y signal to form the luminance signal.

After filtering and the necessary delays, the E'_I and E'_Q signals modulate subcarriers, which are delayed by 57° for the I component and 147° for the Q, with balanced modulators to generate the color

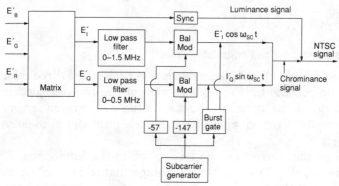

Figure 5.10 NTSC encoder.

subcarrier. The color subcarrier is combined with the Y' luminance signal to form the composite color signal.

The burst is also added by keying the subcarrier with the *burst gate* signal.

The subcarrier can be generated by an atomic clock, generated by a temperature-controlled crystal oscillator, or derived from an incoming signal with a genlock.

5.7.7 Viewfinder

Electronic viewfinders have been a standard feature of studio cameras from their earliest days, and they are functionally superior to the optical viewfinders on film cameras. Viewfinders on modern cameras provide a number of useful features in addition to their basic function:

- Cursor—a remotely controlled cursor provides a means of communication between the director and the camera operator.

- Safety zone—an area encompassing approximately 90 percent of the image marking the useful area of the picture.

- External video connection—a connection that makes it possible for the camera operator to see the output of the camera in relation to other signals. Also, warning and status signals can be displayed.

- Color—viewfinders are now available in color.

5.7.8 Triax cable

The first color cameras were connected to the control unit with three large, heavy cables. With the smaller size, lower power requirements, and greater stability of solid-state components and the use of photo-conductive and CCD imagers, it has been possible to concentrate more of the circuitry in the camera head and to reduce the number of connections between the camera head and the camera control unit. This made it possible to transmit all the video, control, and power connections in a single small cable. The control signals are digitally multiplexed, and the video circuits are transmitted by subcarriers.[3]

Signal	Transmission mode
R video	AM subcarrier, 8.33 MHz
G video	AM subcarrier, 24.98 MHz
B video	AM subcarrier, 41.64 MHz
Sound channels	FM subcarriers, 1.7 and 2.7 Hz
Viewfinder back video	FM subcarrier, 59 MHz
Cue channel	FM subcarrier, 70 MHz
Control channels	PCM

5.8 Camera Operating Controls

Early color cameras had an enormous array of operating controls, and many stations employed a video operator for each camera to keep it in adjustment. The advent of photoconductive storage tubes, CCDs, solid-state circuitry, automatic controls, and general advances in technology have greatly simplified the control problem. Automatic exposure controls that adjust the lens aperture make it possible for the camera to operate over reasonable ranges of scene illumination without adjustment. Digital technology has increased the stability of control circuits so that they do not require frequent adjustment. As a result, hand-held ENG cameras frequently operate without any readjustment of the controls for considerable periods of time. Digital technology has also made it possible to put a number of preset control positions in a memory to be recalled at will.

On the other hand, the ever-increasing demand for the highest possible picture quality has necessitated more sophisticated control of the cameras' adjustments, including precise camera matching. Further, the adjustments may be dictated as much or more by aesthetic judgment as by rigorous technical considerations. The adjustment of the controls by the appearance of the picture rather than by objective test procedures is called *painting*. Needless to say, it requires a carefully adjusted color monitor as the reference.

Table 5.3 shows the operating controls (as opposed to setup controls that would be adjusted during routine maintenance procedures) for a typical studio camera. Function controls that have the option of presetting or automatic operation are indicated. The "painting" category

TABLE 5.3   Camera Operating Controls

	Automatic option	Preset option
Basic picture quality		
Master gain		X
Master black		X
Exposure		
Iris	X	
Color filter		
Neutral-density filter		
Registration		
Centering	X	
Size	X	
Skew	X	
Painting		
White balance	X	X
Black balance	X	X
Flare compensation		X
Shading		X
Aperture response		X
Image enhancement		X
Gamma adjustment		X
Knee adjustment	X	X
Color balance		X

includes adjustments that may be influenced by subjective judgments as well as objective measurements.

5.9   Camera Specifications

Table 5.4 is a tabulation of specifications that are representative of those that the broadcast studios currently offer. Note particularly the notes and the text references.

5.10   Television Film Cameras

5.10.1   A brief history

Before it was possible to record video signals on magnetic tape, film was the only practical medium for the storage of television program material. Television film cameras and their companion projectors were critical components of television station equipment complements, and in many stations most local programming originated on film. The tape recorder, first introduced in 1956, provided an alternative medium, with the additional feature that recordings could be made easily and quickly by individual stations. The advantages of the tape format increased steadily as progress was made in recording

TABLE 5.4 Representative Camera Specifications

Optical system	$f/1.2$, prism
Filters	
Color	3200 K, 5600 K
Neutral-density	¼, ¹⁄₁₆
Sensitivity (Sec. 5.4.8)	
Plumbicon	2000 lux, $f/4.5$
Saticon	2000 lux, $f/4.0$
CCD	2000 lux, $f/5.6$
Minimum illumination (Sec. 5.4.8)	
Plumbicon	10 lux, $f/1.2$, 24-dB gain
Saticon	12 lux, $f/1.2$, 24-dB gain
CCD	15 lux, $f/5.0$, 18-dB gain
Horizontal resolution	
Plumbicon and Saticon	700 lines (center)
CCD	550 lines
Signal-to-noise ratio*	
Phototube	56 dB
CCD	58 dB
Registration†	0.05%

*It is customary to specify the *unweighted* signal-to-noise ratio with gamma correction, aperture correction, and image enhancement removed. This assures that specifications are on a comparable basis. Relative S/N for phototubes and CCD depends on phototube preamplifier.

†This means that the maximum separation of the images for the three colors is 0.05 picture height, or about 0.25 television lines.

technology. As a result, film cameras are now used primarily to make transfers from film to tape. Film, nevertheless, continues to be an important storage medium, and film cameras will continue to be an essential part of the television industry's equipment complement for the foreseeable future.

5.10.2 Frame rate conversion

The conversion of the 24-frame-per-second film frame rate to the 30-frame-per-second rate in countries where this rate is the standard is a basic technical problem. Until the late 1980s, the most commonly used systems were continuous projectors with flying spot scanners in countries that employed 25-frame-per-second television systems and intermittent projectors with storage tube (usually vidicon) cameras in countries where 30-frame-per-second television was the standard.

In 25-frame-per-second countries, the 24-frame-per-second film was simply speeded up. The speed change was usually unnoticeable in the picture, and the difference in sound pitch could be detected only by the relatively small number of listeners who had absolute pitch.

In 30-frame-per-second countries such as the United States, the difference in frame rates was too great to be ignored, and standards conversion was necessary. The intermittent projector and storage tube camera accomplished this. In one such system, the projector had a 3-2 movement—alternate three-field and two-field intervals between pulldowns—to convert the frame rate. The storage capability of the camera made it unnecessary for the exposure and the readout to be simultaneous, and this completed the conversion.

5.10.3 The CCD/continuous projector system

Two major technical developments brought a third system into use that will probably replace the others. It combines a continuous projector, a CCD camera, and a digital standards converter (Sec. 4.5.2).

The scanner generates a digital television signal at the 24-frame-per-second film rate, each frame is stored in a digital memory, and the signal is read out of the memory at the desired television frame rate.

The camera consists of three horizontal rows of CCDs mounted on the red, green, and blue output faces of a prism light splitter. They are precisely located on the prisms so that the images are in registration, and simultaneous R, G, and B signals are generated for each pixel. In combination, the triad of rows generates the signal for one scanning line.

The film frame is illuminated by an incandescent light, and the photographic image is focused on the CCDs. Horizontal scanning is accomplished by a clocked readout of the CCD, and vertical scanning results from the motion of the film image past the CCD.

The functions performed by the electronics for a CCD film camera are similar to those for live cameras, but with some modifications necessitated by the line scan format and standards conversion.

If there is a defect in a single pixel sensor or if its sensitivity varies from that of the others, it will appear as a vertical line in the picture. The fixed pattern correction makes it possible to eliminate or minimize this effect.

The functions of gamma correction, color correction, masking, aperture correction, and image enhancement are performed in the digital mode. The digital signals are reconverted to analog and encoded to composite format or matrixed to component form.

5.11   References

1. Koichi Sadashige, "An Overview of Solid-State Sensor Technology," *SMPTE J.*, February 1987; L. Thorpe, E. Tamura, and T. Iwaski, "New Advances in CCD Imaging," *SMPTE J.*, May 1988. These articles provide an overview of current CCD technology.
2. S. L. Bendell and C. A. Johnson, "Matching the Performance of a New Pick-Up Tube to the TK-47 Camera," *SMPTE J.*, November 1980.
3. D. G. Fink and D. Christiansen, *Electronics Engineers' Handbook*, McGraw-Hill, New York, 1989.

Professional
Video Recorders

6.1 A Brief History of Video Recording

For more than 10 years following the introduction of commercial television at the end of World War II, the problem of developing a satisfactory
magnetic recorder for video signals baffled the industry's major research
laboratories. Such a product was desperately needed, but the technical
difficulties proved too awesome until 1956. In that year the situation
changed dramatically when a small group of young engineers at the
Ampex Corporation discovered a combination of technologies that provided a satisfactory solution called the *quadruplex* recorder.

The key component of the quadruplex recorder was the *headwheel,* a
wheel about 2 in in diameter with four recording heads spaced 90°
apart around the rim. The direction of tape travel was parallel to the
axis of the wheel, and the 2-in-wide tape was partially wrapped
around the rim of the wheel. The tape was pulled past the wheel at a
speed of 15 in/s, and most of the head-to-tape motion was imparted by
the transverse rotation of the wheel at a rate of 240 rev/s. The rotation
resulted in a head-to-tape speed of 1560 in/s, far higher than would be
practical for a longitudinal recorder. The combination of the rapid
transverse motion of the head from the wheel's rotation and the much
slower longitudinal motion of the tape produced a series of recording
tracks, almost normal to the tape, each occupying a time interval of
approximately 11 scanning lines.

Quadruplex recorders were capable of producing excellent pictures
when skillfully maintained and operated, but they were subject to a
number of problems that led to a rapid deterioration of picture quality
if maintenance was less than perfect. The most serious was "banding,"
differences in the appearance of the 11-line bands recorded by adja-

cent heads resulting from small differences in their frequency or phase response. In addition, the quadruplex format lacked flexibility and could not be used for slow-scan or stop-frame applications.

Aided by year-by-year design improvements, the quadruplex recorder had a long life; it was in use for more than 20 years, until the late 1970s. In the meantime, recorders combining the *slant track* or *helical scan* technology with *cassette* tape packaging were enjoying rapid sales growth as the consumer-priced *VCR*, or videocassette recorder.

The helical scan format solved many of the problems of the quadruplex recorder, but it introduced new ones. The most serious was its *time-base instability*, resulting from the long recording tracks on the elastic tape medium. The instability can be tolerated in consumer-level VCRs, but not in complex, high-performance broadcast systems. This problem was eventually solved by the digital time-base error corrector (Sec. 4.5.1), which provided an effective and economical means of stabilizing the time base. (Analog systems were used prior to the advent of digital devices, but it was difficult to achieve the long corrections required for helical scan recording.)

After digital time-base correction was introduced, the replacement of quadruplex by helical scan was almost inevitable, even for the most demanding broadcast and teleproduction applications. The transition was not sudden, owing to the huge libraries of tapes recorded in the quadruplex format, but it was nearly complete by the mid-1980s. As of this writing (1991), helical scan recorders are almost universally used.

The helical scan format is remarkably versatile, and its variations are now used for the entire gamut of video recorder products, from inexpensive mass-produced VCRs to the most complex and sophisticated broadcast and teleproduction systems. This chapter will describe helical scan recorders designed for professional use in the broadcasting and teleproduction industries, while Chap. 7 will describe recorders designed for use in the home.

6.2   The Helical Scan Principle

The helical scan principle is demonstrated by one of the many formats used for helical scan recording, which is illustrated in Fig. 6.1. The tape is wrapped in a helix around a drum on which record/playback heads are mounted, and the drum rotates in a direction opposite to the longitudinal tape motion. The result is a series of diagonal tracks recorded across the tape as shown in Fig. 6.2. In addition to the diagonal video and sync tracks (video and sync are recorded separately in this format), longitudinal recording is used for the audio tracks and for the control track on which the record/playback synchronizing signals are recorded.

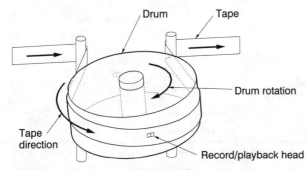

Figure 6.1 Helical scan recorder principle.

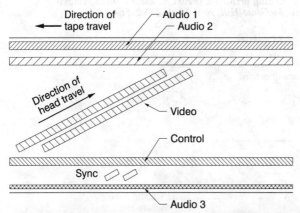

Figure 6.2 SMPTE Type C record format. Ref. SMPTE Type C recording format. (*From SMPTE/ANSI Standards*, V98.18M-1983, V98.19M-1983, V98.20M-1979 (Src), RP85-1979, RP86-1979.)

6.3 The Principle of Magnetic Tape Recording

The basic principle of magnetic tape recording is shown in Fig. 6.3. The signal current passes through a winding around the core of the head assembly, which provides a magnetic circuit for the lines of force that are induced by the signal current. The lines of force, whose density depends on the magnitude of the signal current, fringe into the tape at the low-permeability gap, leaving a pattern of alternating N and S poles on the magnetic surface of the tape. The pattern is called the *record*.

Figure 6.4 shows the detail of the magnetizing force H in the vicinity of the pole-tip gap. The solid curve is the locus of points where H

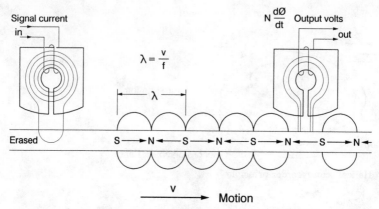

Figure 6.3   Magnetic tape recording principle. (*From K. Blair Benson*, Television Engineering Handbook, *McGraw-Hill, New York, 1986, Fig. 15-1.*)

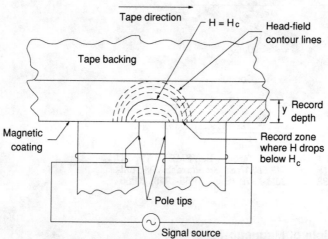

Figure 6.4   Record zone. (*From K. Blair Benson*, Television Engineering Handbook, *McGraw-Hill, New York, 1986, Fig. 15.39.*)

equals the *coercivity* of the magnetic tape coating, H_c. The coercivity is the value of H that is required to magnetize a tape or to return its magnetization to zero. The area within the contour $H = H_c$ is known as the *record zone,* and no magnetism is produced outside this zone. The depth of magnetization and the diameter of the zone is less for high-coercivity tape, and this results in the ability to record shorter wavelengths and higher frequencies.

The reverse process occurs on playback. Magnetic lines of force from the pattern on the tape pass into the core of the head and induce a voltage in the winding that is proportional to the rate of change, the derivative with respect to time, of the magnetic field strength.

$$e = C \frac{dB(t)}{dt} \tag{6.1}$$

where C is a constant and $B(t)$ is the magnetic field strength across the gap.

If $B(t)$ is a sinusoidal function, the voltage is proportional to the frequency, and it increases by 6 dB per octave until the effects described in Sec. 6.7.1 reduce the efficiency of the energy transfer and the response declines with increasing frequency.

6.4 Record/Playback Head Design

Video recording imposes stringent requirements on the performance of the video record/playback heads, and improvement of their design has been an ongoing effort over the past 35 years. In combination with improvements in the tape itself, this has been one of the most important factors contributing to the steady improvement in recorder performance.

Among the requirements that video recording heads must meet are the following:

1. They must be able to operate effectively at very high frequencies—10 MHz and upwards for FM-modulated analog signals and 108 Mbits/s and upwards for PCM-modulated digital signals. Among other constraints, this requires that the gap between pole tips be somewhat less than the wavelength λ of the highest recorded frequency as calculated by Eq. (6.2).

$$\lambda = \frac{\text{head-to-tape writing speed}}{f} \tag{6.2}$$

2. The pole tips must be sufficiently narrow to record and play back on very narrow tracks to provide a high recording density.

3. The head material must have high *permeability*, the ratio of the magnetization flux density B to the magnetizing force H, so that a satisfactory flux density is induced with a modest magnetizing force.

4. They must have a high ohmic resistance to minimize losses from eddy currents.

5. They must be capable of being machined or otherwise formed with very close dimensional tolerances.

6. They must be resistant to wear from abrasion and to mechanical damage from cracking or chipping.

The major research and engineering effort that has been devoted to head materials and designs has resulted in an array of different designs, many of which have proprietary features.

The requirements for pole-tip and core-structure materials are different, and most heads have a composite construction, with one material for the pole tips and another for the core.

Figure 6.5 shows the essential features of a typical composite head design. It has a supporting core structure to which pole tips are bonded. Like the resistance of electric conductors, the reluctance of magnetic conductors is inversely proportional to their cross section, and this is large except in the area of the pole tips. The pole tip is narrow to generate a narrow recorded track.

The core material is usually ferrite, a mixture of iron oxide, Fe_2O_3, and manganese, nickel, or zinc, which has high permeability and resistivity. Ferrites come in two varieties, hot-pressed and single-crystal. Single-crystal ferrites can be machined or otherwise shaped more accurately, but the crystallographic orientation must be carefully controlled.

The pole-tip material has the additional requirement that it must be resistant to wear by abrasion and mechanical damage. This was a problem for ferrites; it was reduced by a glass bonding technique in which a layer of glass is bonded to the surfaces in the vicinity of the

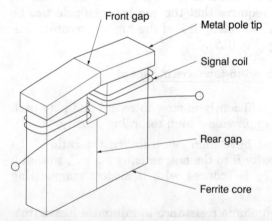

Figure 6.5 Magnetic recording head. (*From K. Blair Benson*, Television Engineering Handbook, *McGraw-Hill, New York, 1986, Fig. 15-43.*)

TABLE 6.1   Properties of Head Materials

Material	Permeability*	Resistivity, Ω/cm†
Alfesil‡	500	90
Ferrite		
Hot-pressed, Mn, Zn	300–500	10^5
Single-crystal	300–500	10^5

*Permeability is flux density B/magnetizing force H; for a complete definition, see Fink and Christiansen, p. 6-65. Permeability is an important specification for head materials because it indicates the flux density and signal voltage generated by the magnetic pattern on playback.

†High resistivity is desirable to minimize eddy current losses.

‡The resistivity of alfesil is increased and its permeability reduced by fabricating from powders.

SOURCES
K. BLAIR BENSON, *TELEVISION ENGINEERING HANDBOOK*, MCGRAW-HILL, NEW YORK, 1986; DONALD G. FINK AND DONALD CHRISTIANSEN, *ELECTRONICS ENGINEERS' HANDBOOK*, 3D ED, MCGRAW-HILL, NEW YORK, 1989.

gap. It also led to the use of iron/silicon/aluminum alloys known by such names as alfesil, alfecon, and sendust. These alloys had the problem of low resistivity, but this was overcome by fabricating them from powders in which the particles were held together by an insulating cement. This reduces the permeability, and a balance must be struck.

The electrical and magnetic properties of representative head materials are tabulated in Table 6.1.

6.5   Tape Transport Configurations

The helical scan format is capable of a variety of tape transport configurations that have been utilized in commercial recorders to meet the varying requirements of different applications.

The tape can be stored on reels for reel-to-reel recording or in cassettes. Cassettes are easier to handle and load, since reel-to-reel machines must be threaded manually, but their capacity is not as great. Capacity limitation has become less of a problem with improved and thinner tapes, and the trend is toward cassettes. All consumer recorders use cassettes because of their ease of loading. The SMPTE Type C recorder, the most widely used professional format, is reel-to-reel, but cassettes are becoming increasingly common as their performance is improved.

Another basic classification is the *wrap* of the tape around the drum. *Omega wrap* (see Fig. 6.1) and *alpha wrap* are named for their resemblance to the Greek letters, and the classification is based on whether the tape path completely surrounds the drum. Most recorders employ omega wrap, although some of these surround the drum except for a very small gap.

Another classification is based on the fraction of a field recorded in one pass of a head, *field per scan* or *segmented scan*.

6.6 Magnetic Recording Tapes

Magnetic recorders originally used a fine iron wire as the recording medium, but this was superseded by tape soon after the tape format was developed by German engineers during World War II. Like magnetic recording heads, magnetic tape has been the subject of an enormous amount of research since the War, not only for its uses in audio and video recording but also for data processing and storage.

A cross section of magnetic recording tape is shown in Fig. 6.6. It consists of a plastic tape base film coated on the front with a magnetic material and on the back with a conductive coating to prevent the buildup of static charges.

6.6.1 Tape base material

The tape base is a polyester chosen for its strength and dimensional stability with respect to temperature, humidity, and time. The base film is chosen to be as thin as possible to minimize the reel diameter, but thick enough for mechanical strength and to prevent print-through between adjacent layers of the magnetic coating on the reel. The smoothness of the surface must be controlled carefully also—as smooth as possible to provide close contact with the recording head, but rough enough to provide adequate friction for tape handling.

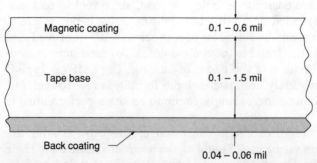

Figure 6.6 Magnetic recording tape. (*From K. Blair Benson,* Television Engineering Handbook, *McGraw-Hill, New York, 1986, Fig. 15-47.*)

6.6.2 Tape magnetic material

The layer of magnetic material is a powder suspended in a binder that bonds it to the base film. The preparation and application of this layer involve processes, often proprietary, that differ from manufacturer to manufacturer.

The desired electromagnetic properties are high permeability and coercivity (see Fig. 6.4). These may seem to be in conflict, since permeability is a measure of the ease with which a material can be magnetized, while coercivity is a measure of the threshold magnetizing force. An ideal material is one in which the threshold is high, so that the depth of the magnetized layer and the minimum wavelength are low, but in which small increases in magnetizing force above the threshold produce large increases in magnetization.

Three magnetic materials, ferric oxide, chromium dioxide (CrO_2), and metal particles, have been commonly used for the magnetic layer. Each has advantages that make it suitable for particular applications. A fourth material, iron oxide doped with cobalt, has desirable magnetic properties but suffers from strong temperature dependence.

The permeability and coercivity of these materials are tabulated in Table 6.2.

Ferric oxide was originally used for magnetic tape and is still the most common material. It is relatively inexpensive and is satisfactory for consumer applications that do not require the ultimate in performance.

Chromium dioxide has another desirable property, a high *saturation magnetization*, the maximum possible value of *B*. With the increased magnetization, a signal-to-noise ratio 5 to 7 dB higher than with ferric oxide can be achieved. A disadvantage is that it is more abrasive.

Metallic particles suspended in lacquer to reduce rusting have high coercivity, high output signal level, high signal-to-noise ratio, and a wider frequency response. This is the common choice for applications with demanding performance requirements.

TABLE 6.2 Magnetic Properties of Tape Materials

Material	Permeability, G	Coercivity, Oe
Ferric oxide	5,000	300–350
Chromium dioxide	6,000	300–700
Metal particles	10,000	1000

SOURCES: J. D. Oetting, "A Comparison of Modulation Techniques for Digital Radio," *IEEE Transactions on Communications*, 23(12), 1979; Alauddin Javed, "Signal Transmission Modes," in A. F. Inglis (ed.), *Electronic Communications Handbook*, Chap. 11, McGraw-Hill, New York, 1988.

The recording characteristics of these materials are different, and they require different head designs for optimum results.

6.7 Tape and Record/Playback Head Performance

This section describes the combined performance of magnetic tape and record/playback heads with respect to frequency response and linearity for analog signals and bit rate for digital—the signal criteria that are directly determined by the tape and head characteristics.

6.7.1 Frequency response and bit rate

Achieving an adequate frequency response and and/or bit rate was one of the most difficult problems faced by the pioneers in videotape recording technology. The bandwidth of the original quadruplex recorder RF channels was limited to 6.5 MHz, an enormous technical feat at the time. *High-band recorders,* introduced in 1964, had a bandwidth of 10.0 MHz, another breakthrough. Present professional analog recorders have bandwidths up to 15.0 MHz and above. (Consumer-type recorders typically have bandwidths of 7 MHz; see Chap. 7.)

The equivalent frequency response, half the bit rate, of digital video recorders is even greater. The SMPTE D-2 standard for composite NTSC signals specifies a $4f_{sc}$ sampling rate with 8-bit words (see Table 4.1). This requires a bit rate of 114.6 Mbits/s, approximately equivalent to a frequency response of 57.3 MHz. The standard sampling rates for component signals are 13.5 Mbits/s for the luminance signal and 6.75 Mbits/s for each of the two color difference signals. The corresponding bit rates are 108 and 54 Mbits/s. Combining these signals serially on a single track would require an equivalent bandwidth of 108 MHz, an impractical rate even with the most modern technology. As a consequence, the luminance and color difference signals are recorded on parallel tracks.

At lower frequencies the frequency response of the tape-head combination increases at the rate of 6 dB per octave, as shown by Eq. (6.1). However, as the frequency is increased further and the wavelength [Eq. (6.3)] approaches the gap width, the other effects described below reduce the response so that it falls off with frequency.

The falloff in frequency response at higher frequencies has four sources:

1. *The finite gap width.* The response from a gap of width d is:

$$e = k\left(\sin \frac{2\pi d}{\lambda}\right)\left(\frac{2\pi d}{\lambda}\right) \qquad (6.3)$$

where k is a normalizing constant and d is the width of the pole-tip gap.

This equation has the (sin (x)/x format that describes many aperture processes. Note that there is a null where $d = \lambda$.

2. *The depth of recording.* The deeper the recording, the wider the record zone and the poorer the response at high frequencies. The response begins to fall off at the point where its wavelength is about one-third the record depth. Since the recording depth is less with high-coercivity tape, its high-frequency response is better.

3. *The spacing between the head and the recording layer.*

4. *Azimuth loss.* This occurs when the gap is not aligned at the same angle with respect to the recording track on playback as when recording. This loss can be fairly large, and the heads must be carefully aligned.

In each of these cases, the high-frequency response is a function of the relationship between the relevant critical dimension and the wavelength.

Figure 6.7 shows the frequency response of a typical head-tape combination.

The increase in high-frequency response that has been achieved in recent years is the cumulative result of technology improvements that affect all of these parameters. Gap widths have been reduced by precision fabrication techniques. Tape writing speed has been increased by moving the head past the tape rather than vice versa. Improved tape materials have increased coercivity and permeability. Improved

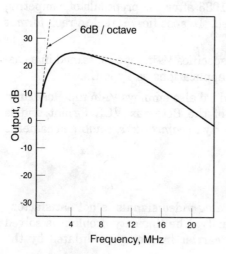

Figure 6.7 Representative frequency response, magnetic recording head.

tape manufacturing techniques have reduced tape roughness and permitted closer head-to-tape spacing.

6.7.2 Linearity

The linearity of the magnetic recording process is determined by the shape of the highly nonlinear B/H magnetization curve. The nonlinearity is so severe that direct recording of a baseband analog audio or video signal is seldom satisfactory. The problem is solved for analog audio signals by adding a high-frequency *bias* waveform to the audio waveform. The problem is solved for analog video signals by the use of frequency modulation. Within reasonable limits, nonlinearity is not a problem for digital signals.

6.8 Analog Recording

6.8.1 Overview

Video signals are analog by nature, and until the late 1970s they were universally recorded in analog form. The attractions of digital recording of analog video signals were numerous, however, and advances in basic recorder technology made digital recording practical as well. Although digital recording has important advantages for some applications and now coexists with analog, analog remains the predominant format except for complex editing systems.

The SMPTE has established three standards for analog helical scan formats with performance satisfactory for professional use:

Type C   This was the original standard for professional helical scan recorders; it was approved in 1983 after the proposals of competing manufacturers were reconciled. It specifies a reel-to-reel format with 1-in tape.

M-II   M-II is a standard that specifies ½-in tape contained in a cassette and is used in the same applications as Type C.

BetaCam   The BetaCam standard also employs ½-in tape loaded in a cassette. It is an upgrade of the Betamax VCR format and is widely used for ENG, particularly in *camcorders,* combined cameras and recorders.

6.8.2 Analog signal modulation

Direct magnetic recording of analog video signals is not satisfactory because of its inherent nonlinearity. The linearity problem is solved by the use of a high-frequency carrier frequency modulated by the video signal. This was introduced by Ampex in its original quadruplex product, and it has continued to be the predominant modulation format for analog signals.

Frequency-modulated video signals differ from FM communication signals in three important respects:

1. The carrier frequency is only a little higher than double the highest-frequency component of the signal.

2. The frequency deviation relative to the width of the baseband is much lower.

3. The dc component (Fig. 1.6) in the signal is retained by setting the frequency representing the sync tip at a fixed value regardless of the content of the signal.

As the result of these differences, the properties of the FM video signals used in recording differ considerably from the FM communications[1] signals. The SMPTE Type C recording standards are an example. The specified instantaneous frequencies in megahertz are

	NTSC	PAL
Peak white	10.0	9.3
Blanking	7.9	7.8
Sync tip	7.06	7.06

The useful picture information is transmitted by a carrier deviation of only 2.1 MHz (7.9 to 10.0 MHz) for NTSC and 1.5 MHz (7.8 to 9.3 MHz) for PAL. The deviation for the complete composite video signal, including sync and blanking, is 2.94 MHz for NTSC and 2.24 MHz for PAL. With a maximum video signal frequency component of 4.2 MHz, the modulation index for the composite NTSC signal is only 0.7. This compares with a typical modulation index of 2.5 for microwave and satellite systems. The small modulation index is required to minimize the lower second-order sidebands $(f_c - 2f_m)$ that would overlap the baseband or even extend into the negative-frequency region (phase reversed, sometimes described as "folded back").

The frequency spectrum of an FM carrier modulated in accordance with SMPTE Type C standards is shown in Fig. 6.8. The bandwidth of early quadruplex recorders was not sufficient to record the upper sidebands, and the result was, in effect, a single- or vestigial-sideband signal. The removal of part or all of one sideband introduces an amplitude-modulated component that can be removed with a limiter. See Sec. 9.4.2.

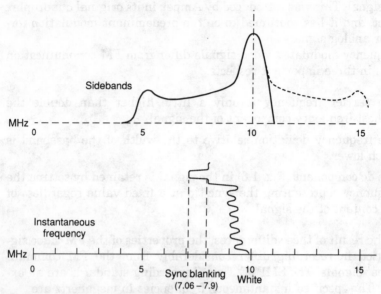

Figure 6.8 Frequency spectrum, Type C record.

6.8.3 Signal-to-noise ratio

The characteristics of FM video signals make the calculation of the signal-to-noise ratio complex and imprecise. The problem is complicated further by the fact that significant noise components can be generated by three sources, the tape, the head, and the electronics. A mathematical analysis of S/N in FM video recording systems is beyond the scope of this volume, but it is important to consider the factors that affect it.

1. The low modulation index adversely affects the S/N in two respects: S, the signal power, is proportional to the square of the modulation index, and the *FM improvement factor* (see Sec. 11.9.2), the ratio of the S/N of the demodulated baseband signal to the S/N of the carrier, is much smaller than in communication systems.

2. The S/N of the demodulated signal decreases linearly with the carrier amplitude down to a critical *threshold*, at which point the S/N drops very rapidly and the signal becomes unusable.

3. As with FM communication systems, the signal-to-noise ratio can be improved by preemphasis of the high-frequency components of the video signal before recording, with complementary deemphasis on playback. The preemphasis standard is CCIR 567 (based on Report 637).

4. Also in common with all FM systems, the amplitude of the frequency components of noise generated by the recording and playback process (not noise in the original baseband signal) rises linearly with frequency. This leads to a greater difference between weighted and unweighted noise than with flat or white noise (see Sec. 2.16.1) because high-frequency components are not as visible. A typical weighting factor is 12 dB.

5. The quality of the tape has a major effect on the S/N. Metal particle tape, with its high coercivity, improves the signal-to-noise ratio by several decibels as compared with conventional iron oxide tape.

The unweighted signal-to-noise ratio currently being achieved on professional helical scan recorders is approximately 49 dB; see Table 6.3.

6.9 SMPTE Type C Recorders

6.9.1 Transport geometry

The SMPTE standard dimensions of the Type C record drum assembly are shown in Fig. 6.9a (plan view) and b (elevation).

The assembly includes a rotating upper drum on which the head pole tips are mounted and a slightly smaller stationary lower drum

TABLE 6.3 SMPTE Recording Standards and Typical Performance, Type C, M-II, and BetaCam Analog Recorders

	Type C	M-II	BetaCam
	Standards		
Tape handling	Reel-to-reel	Cassette	Cassette
Track geometry	Full field	Full field	Full field
		Dual track	Dual track
Tape width, in	1	$\frac{1}{2}$	$\frac{1}{2}$
Tape speed, in/s	22	2.6	4.6
	Typical Performance		
Playing time	126 min, 11.75-in reel	90 min, cassette	90 min, large cassette
S/N ratio, unweighted			
Composite, dB	49	47	47
Luminance, dB		49	49
Bandwidth, MHz			
Composite	4.1	4.2	4.1
Components			
Luminance		4.5	4.1
Chrominance		1.5	1.5
Differential gain, %	<4	<3	<3
Differential phase	<4°	<3°	<3°

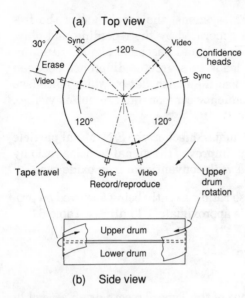

Figure 6.9 Type C record drum. (*From Donald G. Fink and Donald Christiansen,* Electronic Engineers' Handbook, 3d ed., *McGraw-Hill, New York, 1989, Fig. 20-69.*)

that maintains contact with the reference (lower) edge of the tape to provide accurate alignment.

The upper drum makes a complete revolution each field, or approximately 3600 rpm for the NTSC system, and each track records a complete field. This gives the Type C recorder two important advantages over segmented systems: the problem of matching heads is eliminated, and full-quality variable-speed playback is possible (variable-speed operation of a segmented system results in a bar across the picture).

There are three pairs of pole tips on the upper drum, each pair having one tip for video and one for sync. A servo system synchronizes the rotation of the drum with the field rate so that vertical blanking occurs during the short time gap at the end of a video track as the head completes one track and begins the next. The output is switched to the signal from the sync head, which is interrupted at a different time because of its offset location, during this period. Some early recorders used a single head and reconstructed the sync and blanking signals during this gap, but recording the sync signal on a separate track provides more positive synchronization.

The pole-tip pairs have different purposes. The first pair performs the basic record/playback functions. The second pair provides simultaneous playback of the recorded material for monitoring the recording. It is sometimes called the "confidence head." The third pair provides the facility for selective erasure of portions of the record for editing purposes.

6.9.2   Video, audio, and control track records

Figure 6.2 shows the record pattern for the Type C recording standard. The following dimensions should be noted.

The width of the video and sync tracks is only 0.005 in. To maintain this tolerance on playback would be extremely difficult without automatic tracking (see Sec. 6.9.4.2).

The video tracks are more than 16 in long, and the timing errors introduced by an elastic medium of this length necessitate the use of time-base error correction.

The sync tracks are about 16 lines long, sufficient time for the transition of the video pole tip from one track to the next during the vertical blanking interval. The sync signal is switched off during the remainder of the field.

The record track is nearly parallel to the reference edge of the tape, the slant angle is only 2.5°.

6.9.3   Record electronics

Figure 6.10 is a functional diagram of the record electronics for a Type C recorder.

In this design, the video input is ac-coupled, and dc restoration is provided by the feedback clamp circuit that fixes the level of the back porch of the blanking pulse. The gain of the video amplifier is adjusted

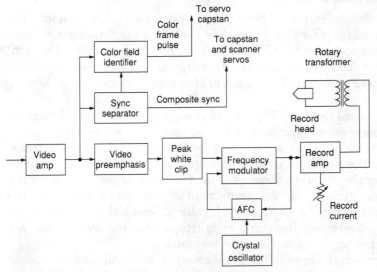

Figure 6.10 Record electronics. (*From K. Blair Benson,* Television Engineering Handbook, *McGraw-Hill, New York, 1986, Fig. 15-60.*)

to set the frequency deviation of the modulated carrier at the values specified in Sec. 6.8.2. The video signal is preemphasized, and signal excursions that exceed the established white level are clipped. The blanking level frequency is established by an automatic frequency control (AFC) circuit, and the modulated carrier is generated by the frequency modulator.

The gain of the record amplifier should be adjusted to bring the magnetizing field H just above the saturation level. This is known as *saturation recording* (see Sec. 6.6). The magnetization curves are different for different pole-tip materials, and the equalization of the record amplifier must be adjusted accordingly. For example, alfesil heads require that the record current be independent of frequency, while ferrites require that the current decrease with increasing frequency. Care must be exercised to avoid the introduction of even-order harmonics of the carrier and its sidebands, which will beat with the first-order sidebands and generate moiré artifacts.

The record electronics must also generate the control signals to the drum servos that synchronize the rotational speed and phase of the drum with the scanning pattern of the recorded signal. The synchronizing signals include the sync pulses and the identification of the first field in the sequence of four for NTSC or eight for PAL; see Secs. 3.14.3 and 3.15.3.

6.9.4 Playback electronics

The electronics employed by playback systems are more complex than those used for recording, since they perform a large number of functions that support and enhance the basic playback function. These include the detection and correction of dropouts, time base and velocity error correction, and slow-scan and stop-motion modes.

6.9.4.1 Signal channel. Figure 6.11 is a simplified functional diagram of the playback electronics of a helical scan playback system.

The head is connected to the preamplifier by means of a transformer. The head/transformer/preamplifier combination has a major effect on the signal-to-noise ratio, and it must be designed carefully for compatibility of the components. The input stage of the preamplifier may be an FET or a bipolar differential amplifier. Its input impedance and the turns ratio (step-up) of the transformer are among the parameters that must be matched.

The preamplifier is followed by an equalizer with rising response at higher frequencies, like an aperture corrector. Typically its response is a constant plus a cosine function beginning at 180° (−1.0) at 0 MHz and continuing to 360° (+1.0) at the upper end of the fre-

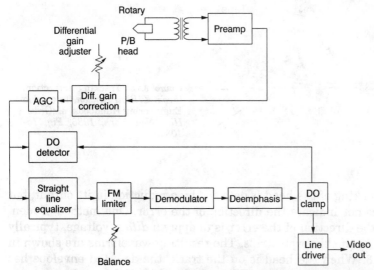

Figure 6.11 Playback electronics. (*From K. Blair Benson,* Television Engineering Handbook, *McGraw-Hill, New York, 1986, Fig. 15-61.*)

quency band. This compensates for the losses in the record/playback process, Fig. 6.10.

The equalizer is followed by an automatic gain control (AGC) amplifier and a dropout (DO) *detector* (see Sec. 6.9.4.5).

The AGC amplifier is also connected to the limiter and the demodulator in the main signal channel. The video signal from the demodulator passes successively through a low-pass filter to remove spurious signal components, a deemphasis circuit to complement the preemphasis in the recorder, and the dropout *corrector*.

The sync and burst waveforms are separated from the output signals and used to synchronize the output of the recorder with the studio sync systems; see Sec. 6.9.4.3.

6.9.4.2 Automatic scan tracking. A means must be provided to cause the playback head to follow the long and narrow recorded tracks. This is accomplished by a technique known as *automatic scan tracking*.

The playback head is mounted on a *bimorph,* two thin flexible strips of a piezoceramic material bonded together; see Fig. 6.12. The strips bend when a voltage is applied across them, and they are mounted so that this causes the head to move at right angles to the recorded track. During the fabrication of the bimorph, the strips are polarized so that a bias voltage is required to maintain them in an undeflected position. Deflection in either direction is achieved by adding the deflection voltage to the bias.

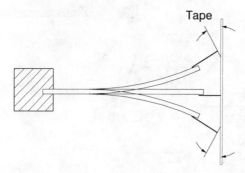

Figure 6.12 Tracking correction. (*From K. Blair Benson,* Television Engineering Handbook, *McGraw-Hill, New York, 1986, Fig. 15-108a.*)

Mistracking of the head is detected by a reduction in its output, but this does not indicate the direction of the error. One method of identifying the direction of the error is to apply a *dither* voltage, typically 500 Hz, to the bimorph strips. The resulting waveforms are shown in Fig. 6.13. When the head is on the track, the detected envelope has double the dither frequency. When the head is off the track, the envelope has a component at the dither frequency; this component is of opposite phase for locations of the head above and below the track. The phase difference is detected by a synchronous detector, and a correction voltage is generated to return the head to the track. (A similar principle is employed for the tracking of laser disks.)

Although it is not necessary to employ scan tracking for playback at normal speeds, it is an additional assurance.

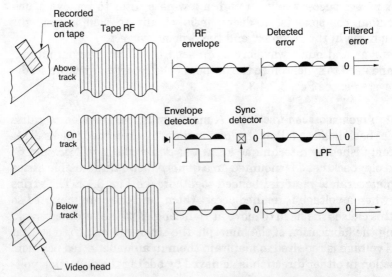

Figure 6.13 Mistracking detection. (*From K. Blair Benson,* Television Engineering Handbook, *McGraw-Hill, New York, 1986, Fig. 15-109.*)

6.9.4.3 Synchronization. It is usually necessary to synchronize the scanning pattern and burst phase of the recorder playback with the studio sync generator so that recorder signals can be combined with other studio signals for mixing and editing. Three events must be timed and synchronized: the beginning of the four-field subcarrier cycle (Sec. 3.14.3), the beginning of each line, and the burst subcarrier phase.

The synchronization of fields and lines is performed by the servos that govern the rotation of the drum and capstan (the shaft that imparts the linear motion of the tape). The servos are controlled by comparison of the timing of signals on the control track with the sync generator output.

The precision required for synchronization of burst frequency and phase is measured in nanoseconds and is orders of magnitude greater than can be achieved by electromechanical means. This is one of the functions of a digital time-base corrector, Sec. 6.9.4.4.

6.9.4.4 Time-base correction. As described earlier, digital time-base correction was a key technology in the development of helical scan recorders for professional use. The operation of time-base correctors is described in Sec. 4.5.1. The time-base correction circuitry may be incorporated in the recorder, or it may be a separate unit.

6.9.4.5 Dropout compensation. The first step in compensating for dropouts is detecting them. The amplitude of the envelope of the FM carrier is continuously monitored, and a drop in level of the order of 16 dB or more produces a pulse lasting for the duration of the dropout. This pulse is then transmitted to the dropout compensator (DOC), which fills the gap in the signal produced by the dropout by interpolating the signal from a corresponding position on an adjacent line. Since the immediately adjacent line is in an alternate field, the phase of its color subcarrier is reversed, and this would cause an objectionable color error. This problem can be solved by interpolating a signal from an adjacent line on the same frame, two lines removed, or by processing the subcarrier from the adjacent line to reverse its phase.

6.9.4.6 Slow-scan and still-frame operation. The helical scan format is particularly adapted to slow-motion and still-frame operation. These effects are achieved by slowing or stopping the tape capstan drive while continuing the drum rotation at its normal speed. Since the angle of the track with respect to the reference edge of the tape is determined by the relationship of the tape and rotational speeds, changing the tape speed automatically causes mistracking.

Figure 6.14 shows the relationship between the record tracks and the playback path under three conditions:

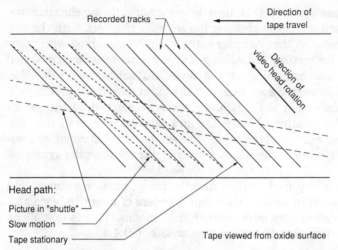

Figure 6.14 Slow motion, still frame, and picture-in-shuttle record tracks.

1. Tape motion stopped; still-frame operation

2. Tape motion slowed moderately; slow-motion operation

3. Tape motion slowed or accelerated greatly; "picture-in-shuttle"

When the tape motion is stopped on playback, the head does not follow the record track perfectly because the track angle has no tape-motion component. In the absence of tracking correction, the end of the playback paths will coincide with the ends of the preceding record tracks. This mistracking can be corrected by an automatic scan tracking system such as that described in Sec. 6.9.4.2.

The mistracking in the slow-scan mode is even smaller, and this, too, can be corrected by automatic scan tracking systems.

The "picture-in-shuttle" mode can be used either to greatly accelerate or to reverse the tape motion. The mistracking is so severe that tracking correction is not possible or practical, but the resulting picture contains enough information to be recognizable for cueing or editing, even when the tape is run backward or moved at 20 or 30 times its normal forward speed. Picture-in-shuttle, therefore, is a useful mode for rapid search of picture content for the location of a particular frame as required for these purposes.

6.10   ½-in Cassette Recorders

Major improvements in tape performance, particularly the introduction of metallized tapes, together with improved heads and circuitry, have made it possible to produce recordings on ½-in tape with a quality level that approaches that of previous recordings on 1-in tape.

When narrower tape became practical for professional use, the logical next step was to convert to the cassette configuration. One-half-inch cassette recorders of professional quality are now widely used, both in portable configurations for electronic news gathering (ENG) and in studio configurations.

Two SMPTE standards, M-II and BetaCam, have been established by the SMPTE for professional-quality ½-in cassette recorders; see Sec. 6.8.1.

Some of the significant record parameters and typical performance specifications of the BetaCam and M-II formats are shown in Table 6.3. The BetaCam and M-II records have parallel dual tracks recorded by closely spaced heads. Both employ component recording with the luminance signal Y recorded on one track, and the color-difference chroma signals $R - Y$ and $B - Y$ on the C track. The components $R - Y$ and $B - Y$ are combined for recording by time-division multiplex.

All signal components employ frequency modulation.

A carrier frequency lower than that used for composite signals can be tolerated for component recording because the absence of a high-energy color subcarrier greatly reduces the crosstalk between the baseband and modulated carrier sidebands. A lower deviation can be tolerated because of technical advances that have improved the signal-to-noise ratio, particularly the use of metal tape.

Figure 6.15 is a drawing of the M-II record format. Note the parallel pairs of Y and C tracks. Also note that, unlike in Type C, the head rotation and tape travel are in the same direction.

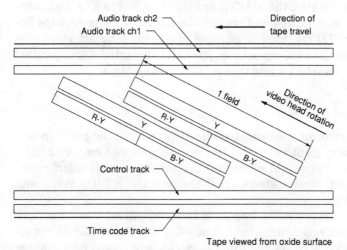

Figure 6.15 M-II record format. (*From Panasonic Corporation*, Instruction Manuals, 17-11 recorders.)

6.11 Digital Recording

The nearly transparent characteristics of the digital video format are particularly important in complex editing systems and other situations where many rerecordings or "generations" are necessary. Its relative freedom from signal-to-noise and linearity degradations as compared with the analog format makes it almost a necessity for these applications.

Its usage was hindered initially by its high bit rate requirements, but this problem has been overcome by steady improvements in tape and head technology. Another problem was the establishment of standards. This has been particularly difficult because of the many choices that are available for digital recording and because of the growing number of broadcasters, manufacturers, and engineering laboratories whose views had to be considered. The problem was addressed seriously in the late 1970s, when representatives of all industry segments began standardization meetings under the aegis of the SMPTE (DTTR) in the United States and the EBU (MAGNUM) in Europe.

6.12 Digital Recording Standardization Policy

Early in the standardization process, it was realized that there were serious conflicts of interest between competing manufacturers with respect to equipment standards. Fortunately, it was possible to resolve many of these conflicts by standardizing the tape and its record rather than the apparatus for achieving it—a major policy change. This assured interchangeability of tape records (the principal purpose of standardization) while retaining the benefits of competition.

In the United States, the SMPTE has led the effort to establish standards for digital recording and created the Digital Television Tape Recording group, DTTR, for the purpose of establishing standards for 525-line systems. At the same time, the EBU appointed a group called MAGNUM to establish standards for 625-line systems.

6.12.1 Standard formats

When digital recording became technically practical, the reaction was to go all the way in picture quality improvement and employ a component format that would eliminate the artifacts, cross-color, cross-luminance, and other defects of the composite NTSC, PAL, and SECAM systems.

The first digital recording format to be standardized was for component signals and was designated D-1 or 4-2-2 (see Table 4.2). The sampling rate is 13.5 MHz for the luminance component and half that, or

6.75 MHz, for each of the two color difference components. The designation 4-2-2 resulted from the fact that a proposed sampling rate for the composite format was four times the subcarrier frequency. This was considered for the luminance channel of the component system, and the nomenclature remained even after a different rate, 13.5 MHz rather than 14.3 MHz, became the standard.

The D-1 format was an excellent choice for internal use in editing systems, but there was also a need for a standard for composite digital signals that could provide an interface with analog systems with a straightforward D-to-A conversion. This became the D-2 format. See Table 4.2.

6.12.2 Scope of standards

Standardizing the tape rather than the apparatus requires detailed specifications to assure interchangeability. For example, the D-1 format is defined by five SMPTE standards:

SMPTE 224	Tape Record
SMPTE 225	Principal Properties of Magnetic Tape
SMPTE 226	Dimensions of Tape Cassettes
SMPTE 227	Signal Content of Helical Data Records of Associated Control Records
SMPTE 228	Signal Content of Cue and Time Code Longitudinal Records

6.13 The D-1 Format

Important electrical and mechanical standards and a representative performance of the D-1 format are shown in Table 6.4. The tape width is ¾ in, and it is cassette-mounted. The sampling rate is 13.5 MHz for the Y or luminance signal. (See Sec. 4.7.1 for considerations involved in the choice of this rate.) The bandwidth and signal-to-noise ratio exceed the performance of analog formats. The absence of cross-color, cross-luminance, and color artifacts that is characteristic of the component format is an important feature.

The tape record pattern is shown in Fig. 6.16. Each track consists of two video sectors, each having a length of 77.71 mm, separated by a group of four audio sectors, one for each audio channel, each having a length of 2.55 mm. The total length of each track is 170 mm.

The audio segments are located in the center of the tracks because the center of the tape is less prone to dimensional changes and physical damage. The performance of the audio system is excellent. The modulation is digital with a 48-kHz sampling frequency. Quantization is 16 bits per sample, and the result is frequency response to 20 kHz,

TABLE 6.4 SMPTE D-1 Digital Recording Standards and
Typical Performance

Standard	
Video format	Component
Tape width	¾ in
Cassette	D-1 standard
Tape speed	11.2 in/s
Writing speed	
Sampling rate	
Luminance	13.5 MHz
Color difference	6.75 MHz
Typical Performance	
Cassette record time	94 min (L) cassette
Video bandwidth	
Luminance	6.0 MHz
Color difference	2.75 MHz
S/N, each channel (unweighted)	56 dB
Cross-color distortion	None
Cross-luminance distortion	None
Color artifacts	Minor

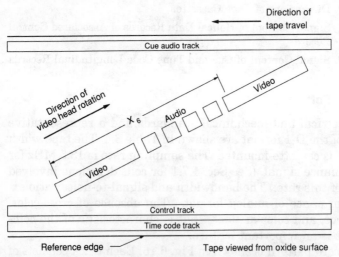

Figure 6.16 D-1 record pattern. (*From Proposed American National Standard for Component Video Recording*, 19mm type D-1 cassette, Helical Data and Control Recorder, SMPTE 227M.)

less than 0.05 percent distortion, and a 90-dB dynamic range. Noise is virtually unmeasurable.

The remaining 25 lines are unrecorded. In 625-line systems, 600 lines are recorded, 60 on each segment. The luminance and color difference signals could be recorded sequentially, but their order is changed or *shuffled* continuously according to a pattern standardized in SMPTE 227. Shuffling is a basic component of the error concealment process; see Sec. 6.15.

6.14 The D-2 Format

The sampling rates that have been used for composite signal transmission are $3f_{sc}$ and $4f_{sc}$, where f_{sc} is the color subcarrier frequency. $4f_{sc}$ or 14.4 MHz gives superior performance and is specified in the SMPTE standard.

Important recording standards and typical operating performance for the D-2 standard are shown in Table 6.5.

6.15 Error Correction and Concealment

Error correction and concealment are of critical importance in digital recording, since its ability to make multigenerational copies of high quality is one of its most important advantages.

The error detection and correction techniques described in Sec. 4.15 are applicable to digital recording. In addition, shuffling, as described

TABLE 6.5 SMPTE D-2 Digital Recording Standards and Typical Performance

Standard	
Video format	Composite
Tape width	¾ in
Cassette	D-1 standard
Tape speed	5.2 in/s
Sampling frequency ($4f_{sc}$)	14.3 MHz
Quantization	8 bits/sample
Typical Performance	
Cassette record time	94 min (M) cassette
Video bandwidth	6 MHz
S/N ratio (unweighted)	54 dB
Differential gain	<2%
Differential phase	<1°
Cross-color distortion	Yes
Cross-luminance distortion	Yes
Color artifacts	Yes

in Sec. 6.13, can be used for *error concealment*. This does not correct errors but makes them less visible.

The action of shuffling in error concealment can be explained as follows. Assume a dropout on the tape that is one pixel high and four pixels wide. Without shuffling, it will produce an image defect with these dimensions. With shuffling, however, a defect one pixel wide, and therefore much less visible, will appear in a vertically staggered pattern on four adjacent lines. It will not have disappeared, but it is much less visible because of the smaller dimensions of the defect on each line.

6.16   Reference

1. M. Felix and H. Walsh, "FM Systems of Exceptional Bandwidth," *Proc. IEEE*, 112(9), September 1965.

Home Video Recorders (VCRs)

7.1 Introduction

The home video recorder, more popularly known as the VCR or videocassette recorder, is one of the great technical achievements of the last two decades. The helical scan principle, the technology on which it is based, was originally developed for audio-video and industrial recorders—higher priced and capable of producing better-quality pictures than expected for consumer products, but inadequate for professional use.

Product development based on the helical scan principle then proceeded in two directions. One path led to recorders of professional quality that surpassed those using the original quadruplex format (see Chap. 6). The other led to VCRs for the home—low-cost, mass-produced recorders that produce pictures of remarkably good quality, although short of professional standards. The combination of performance and cost that is available from today's home VCRs would have been almost unthinkable 20 years ago. The seemingly simple VCR, operating in a home environment, employs technology of a sophistication that rivals that of a professional recorder.

Although professional recorders are available in both reel-to-reel and cassette versions, the term *VCR* is usually used to denote consumer-type recorders for the home, and it is has this meaning in this volume.

7.2 Comparison with Professional Recorders

Although professional recorders and VCRs employ the same basic technology, there are large differences in their design priorities, which result in wide differences in their cost, performance, and operating features.

Features that are given the highest priority for VCRs include

1. Low price
2. Size and weight
3. Recording time and tape usage
4. Simplicity of operation
5. Cassette tape storage

Picture quality, while obviously important for VCRs, is not as critical as for professional recorders.

Features that are given the highest priority for professional recorders include

1. Basic picture quality—bandwidth, S/N, linearity
2. Time-base stability
3. Multigenerational performance
4. System integrability
5. Compliance with FCC B/C standards

The effect of these priorities on product design is shown in Table 7.1, which compares the specifications of representative VCRs and professional recorders. These specifications are chosen to illustrate the significant differences between the two product types; they are not based on specific designs.

TABLE 7.1 Typical VCR and Professional Cassette Recorder Specifications

	VCR	Professional
Price	$500	>$30,000
Size	4 in × 14 in × 12 in	10 in × 17 in × 21 in
Weight	<25 lb	60 lb
Tape width	½ in or 8 mm	½ in or 8 mm
Operator skill	Same as TV receiver	Technician
Recording time	See Table 7.2	
Bandwidth (luminance)	2.5–4.0 MHz	4.3 MHz
S/N (composite)	45–47 dB	49 dB
Differential phase	10°	3°
Time-base instability	>1 μs	2 ns
No. of generations		
Analog*	2	4
Digital	NA	30
System integrability	Poor	Good
Meets FCC B/C standards	No	Yes

*Marginal performance.

7.2.1 Price, size, and weight

Price, size, and weight—particularly price—are perhaps the most critical specifications for recorders intended for the consumer market. They are also the parameters that display the greatest differences between VCRs and professional products.

The size and weight of rack- or cabinet-mounted professional recorders are inappropriate for a home environment. Home recorders are typically of a size that can be mounted on a TV set, on a bookcase, or on a small table and weigh less than 25 pounds.

7.2.2 Operator skill and operational simplicity

Although operational simplicity is an important feature of both professional recorders and VCRs, it can reasonably be assumed that professional products will be operated by trained technicians. VCRs, on the other hand, must perform satisfactorily when operated by completely unskilled individuals. This establishes a requirement for a minimum of adjustments and highly automatic operation.

7.2.3 Playing time and tape storage

Long playing time and cassette-loaded tape are almost absolute requirements for home recorders.

The *VHS* VCR format won a clear victory over the *Beta* format (see Sec. 7.4) in the marketplace because of its longer playing time. Cassette loading is preferred because of its convenience and ease of operation. The playing time of most VCRs is at least 4 hours; the playing time of professional recorders is less because of their higher tape speed—typically 2 hours for reel-to-reel models and 1 to 1.5 hours for cassettes.

7.2.4 Bandwidth, signal-to-noise ratio, and linearity

For direct recording of composite NTSC signals, the bandwidth of the record/playback process must be sufficient to transmit the 3.58-MHz color subcarrier and its sidebands; see Sec. 3.14.4 and Fig. 3.10. The nominal bandwidth requirement is 4.25 MHz.

The signals recorded on the tape by VCRs are not NTSC but another format, known as *color-under* (see Sec. 7.7), in which the chroma subcarrier frequency is about 700 kHz. The color-under signal is reconverted to the NTSC format during the playback process, but its chroma bandwidth is limited by the bandwidth of the color-under subcarrier. The narrow chroma bandwidth is adequate for most picture content, but it can cause color smearing of images with high color saturation and sharp hue discontinuities.

The bandwidth of the luminance channel is a design choice based on a cost/performance tradeoff—one of the many that must be made in VCR design. A bandwidth of 2.5 MHz, the lower limit of the bandwidth range shown in Table 7.1 and a common choice, results in a horizontal limiting resolution of slightly more than 200 lines, not a high-quality picture but one that is satisfactory to many viewers.

The signal-to-noise ratio and the linearity (as indicated by differential phase in Table 7.1) of VCRs also involve cost/performance tradeoffs. The color-under format may result in coarse, low-frequency noise in areas of high saturation.

7.2.5   Time-base stability

As noted in Sec. 6.1, time-base instability is an inherent problem with helical scan recorders because of the long recording tracks and the elastic recording medium. Without time-base correction (TBC), the instability may exceed ±1 µs, several orders of magnitude greater than the ±2-ns instability of professional recorders. Stability is restored in professional recorders with digital TBC (see Sec. 6.9.4.4), but this is too costly for consumer VCRs. A simpler analog VCR TBC is shown in Fig. 7.9.

Most receivers can track the luminance channel of a VCR signal in spite of the instability of its time base, but the instability of the chroma channel, as recorded on a VCR, would cause unacceptable chroma distortion, even by consumer product standards, if it were demodulated with a fixed-phase subcarrier as the reference. This problem can be solved with results that are satisfactory for many applications by controlling the phase of the reference subcarrier with the burst. For professional recorders, a common fixed-phase reference subcarrier must be used for all signal sources; this requires the use of digital or very high-performance analog time-base correctors. VCR signals without time-base correction are sufficiently stable for direct viewing, but they cannot be edited except for the simplest switching operations, and the switched signals are subject to rolls and transients in the chroma pattern at the time of the switch.

The time-base instability of VCRs also makes it difficult or impractical to integrate them in systems of even moderate complexity.

7.2.6   Compliance with FCC broadcast standards

VCR signals generally do not comply with FCC broadcast standards. Their greatest defect is their timing instability, but they may fail to meet FCC standards in other respects, such as sync and blanking level format and signal and blanking levels. They can usually be upgraded, however, to meet FCC standards by the use of digital time-base correctors and other signal processing equipment.

7.3   Recording Drum Configurations

Two recording drum configurations, illustrated in Fig. 7.1, are commonly used in VCR tape transport systems, the essential difference being their diameters. The diameter of the larger drum is 50 percent greater than that of the smaller, and it has two heads rather than four. With these diameters, the length of a 270° track on the smaller drum is equal to the length of a 180° track on the larger. The rotational speeds are controlled so that the smaller drum rotates 270° and the larger 180° during one television field, thus creating record tracks of equal length for one field.

The scanning sequence of the two formats is shown in Fig. 7.2. The configurations are compatible, and tapes recorded on one can be played back on the other.

7.4   VCR Record Formats

Helical scanning is a very versatile technology, and it has been adapted to a variety of formats to meet the widely different requirements of the consumer, audiovisual, and professional markets.

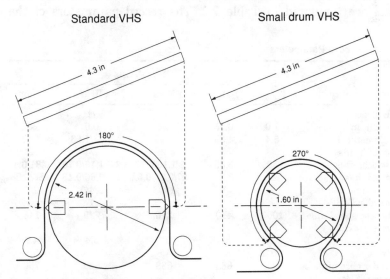

Figure 7.1   Recording drum configurations.

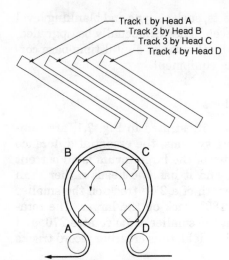

Figure 7.2 Recording track sequence.

A description of all the many VCR formats and their variations (some of which are in a continuous process of change) is beyond the scope of this volume. The formats described here are the most commonly used consumer formats, including the Beta format, which has largely been replaced by VHS.

7.4.1 Format parameters

The parameters of the most commonly used consumer and audiovisual formats are tabulated in Table 7.2. The record parameters of the

TABLE 7.2   Record Parameters

Commercial type		VCRs			
		U-Matic	Beta	VHS	8-mm
SMPTE Type	C	E	G	H	
Tape width, in	1.0	0.75	0.5	0.5	0.31
Drum diameter, in	5.4	4.33	2.95	2.44/1.63	1.57
Number of heads		2	2	2/4	2
Track width, mils	50	3.35	2.3/1.15/0.77	2.3/1.15/0.76	0.81/0.4
Tape speed, in/s	22	3.75	1.57/0.79/0.52	1.3/0.66/0.43	0.56/0.28
Recording time, h		1	1/2/3	2/4/6	2/4
Camcorder (VHS-C)				1	
Head-to-tape speed, in/s	1000	410	276	229	148
FM carrier, MHz					
Sync	7.06	3.8	3.5	3.4	4.2
White	10.0	5.4	4.8	4.4	5.4
Color-under chrominance subcarrier, kHz	NA	688	688	629	750

SMPTE Type C professional reel-to-reel format are shown for comparison. As indicated by their generic title, all consumer formats employ cassettes, and the trend is toward cassettes in professional recorders as well; see Chap. 6.

The U-Matic was the first helical scan format to achieve commercial success. It was an audiovisual and industrial product with the characteristics described in Sec. 7.1, and it is still used in these markets.

U-Matic recorders were followed by a series of lower-priced models designed specifically for the consumer market:

Beta	1976
VHS	1978
8-mm	1983

In spite of its earlier introduction and excellent performance, the Beta format lost the competitive battle with VHS in the consumer market, largely because of the latter's longer playing time.

VHS (including its special-purpose variants) is now the most widely used VCR format, but it, in turn, is receiving competition from 8 mm, particularly for camcorders, where small size is an important advantage.

The Beta, VHS, and 8-mm formats offer a choice of *tape speeds,* usually described as standard, LP (long play), and SLP (super long play). Note especially that the *tape-to-head speed remains constant for different tape speeds,* and tape-speed variations are achieved by altering the track width.

Upgraded (and more costly) models of the Beta and VHS formats have been developed—BetaCam SP or Super-Beta and Super VHS or SVHS—with performance specifications that are adequate for some professional applications.

7.4.2 Azimuth recording

Azimuth recording is incorporated in almost all VCR record formats because of its effectiveness in minimizing the effects of mistracking, reducing crosstalk between adjacent tracks, and improving the S/N. It offers fewer advantages for professional recorders that employ wider tapes, and it does not comply with the standard Type C format.

The record pattern for azimuth recording is shown in Fig. 7.3. The gap in the record head is tilted from its normal 90° orientation with respect to the head tracks, and the magnetic lines of force in the record are not parallel with it. The gap azimuth angle, the direction of tilt, is reversed on alternate scans, and on playback the signal from adjacent scans is attenuated by the difference between the head gap and the magnetic record azimuths.

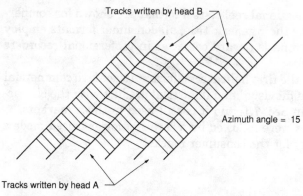

Tracks written by head B

Azimuth angle = 15

Tracks written by head A

Figure 7.3   Azimuth recording—record pattern. (*From Donald G. Fink and Donald Christiansen,* Electronic Engineers' Handbook, 3d ed., *McGraw-Hill, New York, 1989, Fig. 20-74.*)

The precise azimuth angle is not critical, but it is normally set at about ±10° or ±15° from a 90° angle with the head path.

Since signal pickup from adjacent tracks is small, it is unnecessary to provide guard bands on either side of the recorded tracks. This permits the use of wider heads that produce a higher S/N and are less sensitive to tracking errors.

7.4.3   Audio recording

Audio signals can be recorded by a 1.5-MHz (1.4 and 1.6 MHz for stereo) FM carrier, either with separate fixed heads on longitudinal tracks on the edge of the tape or with rotating heads mounted on the same drum as the video; see Fig. 7.4.

In the rotating head configuration, the audio head gaps are located 60° around the periphery of the drum from the video, and their azimuths are tilted ±30° from normal.

The video and audio records occupy approximately the same area on the tape except for additional space on the end of each track that is necessary because of the separation of the video and audio heads; see Fig. 7.4. The signals are kept separate by

1. The difference in carrier frequencies.
2. The difference in the head azimuths.
3. The difference in the strength of the record magnetizing force. The longer gap length of the audio head causes the recording field to penetrate more deeply into the tape. The audio signal is recorded first and the video record is overlaid on it.

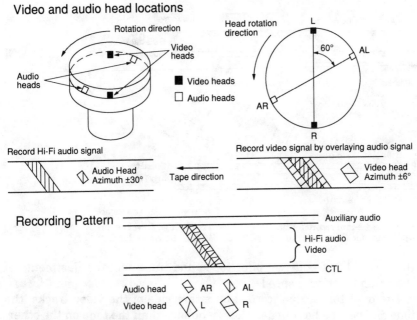

Figure 7.4 Audio head locations.

7.4.4 Format patterns

The format patterns for VCRs are similar in principle to those shown in Chap. 6, e.g., Fig. 6.2, for professional recorders. The diagonal video and audio tracks occupy the center portion of the tape, one field per track, and the fixed-head audio and control tracks are recorded longitudinally along the edges. The 8-mm format, Fig 7.5, shows the essential features of VCR record patterns, although they have many variations.

7.4.5 Record frequency spectrum

The frequency spectrum of an 8-mm record is shown in Fig. 7.6. The frequency-modulated luminance channel occupies the region from 4.28 to 6.8 MHz (including sidebands). The audio signal occupies a band around 1.5 MHz, and the color-under chroma carrier (see Sec. 7.7) occupies a band around 750 MHz. There may be, in addition, four low-frequency pilot tones that are used for automatic scan tracking; see Sec. 7.5.

7.5 Automatic Scan Tracking

Automatic scan tracking (AST) systems that employ bimorph mounting of the heads, as used in professional recorders, were described in Sec. 6.9.4.2. VCRs may use different systems, one of which is based on

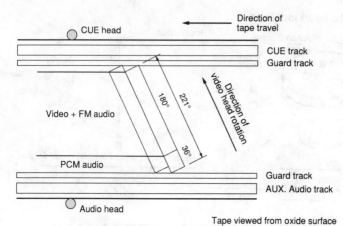

Figure 7.5   8-mm record format. (*From Koichi Sadashig:*, Developmental Trench for Future Consumer VCRs, *J. SMPTE, Decenber 1984.*)

the use of low-frequency pilot tones and incremental adjustments of the drum rotational speed.

The pilot tones are recorded on either side of the video tracks, the tone on one side having a higher frequency than the tone on the other. The head is wide enough to pick up both tones on playback; if it is properly centered, the amplitude of both will be equal. If the tracking is not perfect, the amplitude of one of the tones will be greater, and the capstan servo that determines the tape speed is adjusted by a feedback circuit to restore the tracking.

7.6   Record Information Density

7.6.1   Information density requirements

The VCR marketplace has imposed two conflicting requirements on VCR designs—small size and long playing time. These needs can be satisfied simultaneously only by increasing the *record information density,* i.e., the amount of information recorded per unit area of tape.

The increase in this parameter in the more recently introduced formats shown in Table 7.2 has been dramatic. The area required to record one frame of an NTSC signal with these formats as compared with the SMPTE Type C format is as follows:

Format	Area/frame, in^2 × 100
SMPTE Type C	73
U-Matic	6.25
Beta	2.62
VHS	2.16
8-mm	0.58

Some of the area reduction is possible because of differences in the bandwidths of the recorded signals. For example, the highest-frequency component of the 8-mm RF signal is about 7 MHz, as compared with about 12 MHz for the Type C signal. But most of the reduction of the area per frame is due to higher *information density* on the record pattern.

7.6.2 High-density technologies

The information density of the recorded signal is determined by the wavelength of its highest-frequency component and the *pitch* or spacing of the recording tracks. The track pitch and minimum wavelength of the common VCR formats are:

Format	Track pitch, mils	Minimum wavelength, μm
SMPTE C	50	2.12 (12 MHz)
U-Matic	3.35	1.74 (6 MHz)
Beta	2.3/1.15/0.77	1.40 (5 MHz)
VHS	2.3/1.15/0.76	1.16 (5 MHz)
8-mm	0.81/0.4	0.63 (6 MHz) (metal tape)

Reducing the track pitch and minimum wavelength in the SMPTE Type C format to the values used in the Beta, VHS, and 8-mm standards was a major engineering challenge, and it required the addition or improvement of a number of technologies.

The problems included:

1. Reduction in bandwidth and a lower S/N because of the necessity of narrower recording heads

2. Tracking difficulties because of the narrower tracks

3. Crosstalk between adjacent tracks

The solutions included:

1. Improvements in tape/head performance (see Secs. 6.7 and 6.8). The track pitch and the wavelengths in the 8-mm format are so small that metal-particle or evaporated-metal tape is almost a necessity.

2. Azimuth recording (see Sec. 7.4.2). This reduces the effects of mistracking, permits the use of wider recording heads for improved S/N, and reduces crosstalk between tracks.

3. Automatic scan tracking, AST (see Sec. 7.5).

7.7 Color-Under Signals

7.7.1 Record frequency spectrum and electronics

The principle of the color-under signal format is illustrated in Fig. 7.6, which shows the frequency spectrum of the 8mm record. The essential elements of the record electronics are shown in Fig. 7.7. The chroma component (the color subcarrier) of the incoming signal is separated by a bandpass filter, and the NTSC chroma and luminance components are processed separately. The frequency of the chroma component is down-converted to the 650–750 kHz region by heterodyning and its bandwidth reduced (see Table 7.2) to form the color-under component of the recorded signal. This places the color subcarrier and its sidebands *below* the band occupied by the FM video signal rather than being interleaved with it as with NTSC and PAL; hence the term *color-under*. Since the bandwidth available for the color-under signal is limited, it is necessary to pass it through a band-limiting filter.

The luminance component passes through a low-pass filter with a bandwidth of about 2.5 MHz (the luminance component has a wider bandwidth in some "high-band" models), and it frequency-modulates the video carrier between the limits shown in Table 7.2.

The audio signal is recorded on a track along the edge of the tape, or alternatively it is multiplexed with the video record; see Sec. 7.4.3.

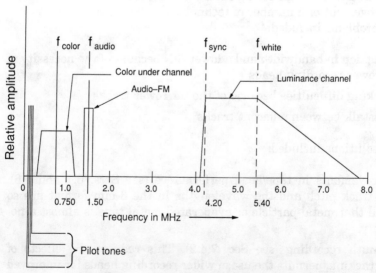

Figure 7.6 Record frequency spectrum, color-under signal.

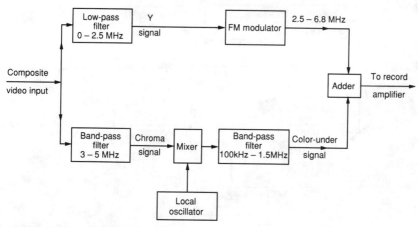

Figure 7.7 Record electronics.

If pilot tones are used for automatic scan tracking, they are recorded at the bottom of the spectrum, as shown in Fig. 7.6.

7.7.2 Playback electronics

Unlike professional recorders, which require a time-base stability of about ±2 ns on playback so that the signal can be mixed and edited with other signals (see Chap. 8), VCRs, which are not used in complex editing systems, can tolerate a considerably higher degree of instability. The use of time-base correctors is an option that can be included, however.

Figure 7.8 is a functional diagram of the electronics of a typical playback system. The chroma and FM luminance signals are separated, and the luminance signal is demodulated. As noted in Sec. 7.2.5, the time-base instability of the chroma signal is too great to operate satisfactorily with a fixed reference subcarrier, and the reference subcarrier is generated by a burst-controlled oscillator whose phase varies with the timing errors on the tape. This retains the phase relationship between the color and reference subcarriers to a reasonably satisfactory degree.

Improved performance can be obtained with a relatively inexpensive analog time-base corrector; an example is shown in Fig. 7.9. The playback video is separated by filters into its chroma and luminance components. The luminance component, its bandwidth limited to about 2.5 MHz, passes directly to the output. A burst-controlled subcarrier is generated, as in the playback electronics in Fig. 7.8. This

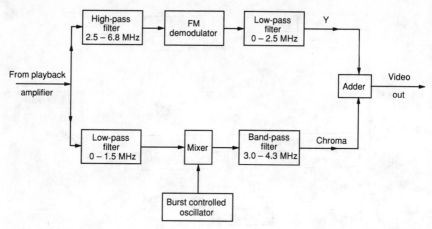

Figure 7.8 Playback electronics.

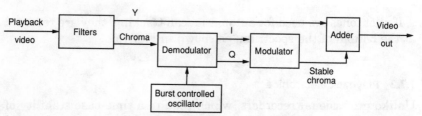

Figure 7.9 VCR time-base corrector. (*From K. Blair Benson,* Television Engineering Handbook, *McGraw-Hill, New York, 1986, Fig. 15-65.*)

subcarrier is then used to demodulate the *I* and *Q* or color difference signals. These, in turn, remodulate a stable subcarrier, which is combined with the uncorrected luminance signal. This gives a stable chroma signal combined with an unstable luminance, a result which is satisfactory for many applications.

8

Editing Systems

8.1 Historical Overview

Television editing is the assembly and revision of program material by adding, deleting, modifying, and combining signals from one or more sources. Before magnetic tape recording was available, the editing process employed film editing techniques with kinescope recordings or was limited to live production. Kinescope recordings did not receive wide acceptance because of their marginal quality, and most program editing was done live, employing multicamera techniques. For dramatic programs of any complexity, this required careful preplanning and rehearsals, not unlike those for a stage play, and editing was accomplished in real time. It was fast, but if a mistake was made, there was no second chance.

In spite of the drawbacks of live production, it had certain advantages over the methods used to produce and edit film, which involve shooting a series of very short scenes or takes with a single camera, one take at a time. The takes are then selected, edited, and assembled by a laborious manual process. Single-camera film editing in this manner has almost unlimited flexibility, and repeated changes can be made, but it is slow and costly, and attempts were made to adapt multicamera television editing techniques to film making.

Multiple film cameras are equipped with television-camera viewfinders so that the camera operators and director can see exactly what is being exposed on the film, including the effects of zoom lenses and camera switching. This adaptation of television editing technology to film making has been only moderately successful because of technical and labor relations problems. It is rarely, if ever, used for the production of theatrical films, and it is not widely used for television film series.

By contrast, there has been a great demand for a means of editing magnetic tape that would have the same flexibility as film editing while retaining the advantages of multicamera television editing. Many technical problems were encountered, and they delayed the full development of this technology for many years. One by one the technical problems were solved, however, and today's tape editing systems are highly developed and widely used.

8.2 Basic Technical Advances

Two basic technical advances were necessary for the development of a successful tape editing technology. The first was the ability to make multiple-generation copies (i.e., copies of copies of copies) of tape recordings with only a minor loss in image quality; the second was the ability to maintain very precise synchronization between signals from different sources.

After these technical advances were achieved, it became possible to perform complex computer-controlled editing processes with tape recordings that equal or exceed the capability of film editing processes.

8.2.1 Multiple-generation copies

The outputs of complex editing programs are almost invariably multiple-generation copies that have been produced by successive recording and rerecording of original recordings or live material. With analog recording, this imposes a sometimes impossibly severe requirement on the quality of each generation, since degradations in signal-to-noise ratios are cumulative.

In the earliest days of video recording, beginning in 1956, the quality of even first-generation recordings was poor by today's standards. Slowly but steadily, quality improvements were made, climaxed by the high-band technical breakthrough in 1964 (see Chap. 3) that made second-generation (copies of originals) or even third-generation (copies of copies) recordings usable, although their quality was far from ideal. After 1964, incremental but steady improvements in high-band recording continued, but the final breakthrough in image quality for multiple-generation systems came in the late 1970s with the introduction of digital recording. Within limits, successive generations of digital recordings suffer virtually no loss of quality (see Sec. 4.3.2); consequently, the digital format is ideally suited for advanced editing systems.

8.2.2 Field, line, and subcarrier synchronization

Nearly all editing processes require signals from different sources to be combined, and synchronism between these sources must be maintained in the edited output. Synchronization can be maintained at three levels of precision: field by field, line by line, and subcarrier phase.

In the absence of field-by-field synchronization, the picture will roll as it is switched from one source to another. Field-by-field synchronization does not have to be particularly precise, and it requires only that the field rates be equal and that the starting time of the fields be synchronized within a few lines.

Before third-generation copies became practical, editing was accomplished by carefully cutting tape segments with a razor blade and splicing them together. This technique, in addition to being crude and time-consuming, could not achieve more than field-by-field precision.

Line-by-line synchronization requires that the starting times for corresponding lines on two or more sources be matched within a few microseconds, preferably within a microsecond. Even this precision is not adequate for color signals, which require subcarrier synchronization except for simple switches.

Subcarrier phase synchronization is required for color signals when it is necessary to mix two or more signals during the same line, either by superposition or by switching. To be effective, phase synchronization must be maintained within very close limits, ± a few degrees at the subcarrier frequency.

8.3 Basic Editing Functions

The technology and techniques of editing have progressed from simple additions and deletions to computer-controlled functions producing complex visual effects that can be duplicated only with difficulty, if at all, in film systems. Advanced technologies have been developed for this purpose that are beyond the scope of this volume; see the references. This section is devoted to a description of basic editing functions followed by a summary of the operation of more advanced systems.

8.3.1 Copy additions and deletions

The most basic editing function is the equivalent of a hard-copy editor with a blue pencil. Assuming that the copy is created by a modern computer program, this type of editing is straightforward, whether the copy is alphanumeric characters or the product of a more complex computer program, such as a graph. It is the most basic editing function, and the relative ease with which it can be accomplished is one of the important features of computer-generated copy.

8.3.2 Switching

Switching from program source to program source is also a fundamental editing function. The obvious technical requirements are that the amplitudes of the sources be matched and that the field-beginning times be closely synchronized (see Sec. 8.2.2). Receivers can recover quickly with little visual effect from small discontinuities in line starting times or subcarrier phase.

8.4   Mixing and Effects (M/E)

In more advanced editing systems, signal combining functions of greater complexity, *mixing* and *effects,* are utilized in addition to simple switching in making transitions from one program source to another.

Mixing is a term borrowed from audio technology and describes a function in which two or more signal sources are directly added or mixed. The amplitudes of the signals can be automatically adjusted so that the sum is held constant at the single-signal level. Switching between signals is accomplished by reducing (fading) the amplitude of one while simultaneously increasing that of the other, an operation called a *lap dissolve.*

Effects describes functions in which source switching occurs during a frame or line. Unlike mixing, where the output is the sum of two or more signals in some proportion, effects produce images that at any single point are one input signal or the other. The difference is illustrated in Fig. 8.1.

The output of an effects system is a composite of visual images, either natural or artificial, in an arrangement that is instructive, informative, or aesthetically pleasing.

The advent of the digital mode, in which millions of individual pixels can be easily manipulated, greatly increased the versatility of the editing process, not only for editing but also for the creation of artificial images. Editing today is far different from the process in the late 1950s, when it was accomplished with a razor blade (see Sec. 8.2.2)!

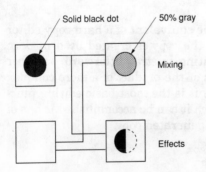

Solid black dot   50% gray

Mixing

Effects

Figure 8.1 Mixing and effects.

Mixing and Effects

8.5   Effects Technologies

8.5.1   Electronic boundary generation

From the earliest days of television, the boundaries of the effects patterns, which define the points on each line where switching occurs, have been generated electronically. *Keying signals* switch the output from one input signal to another at the boundaries of the pattern. Prior to the introduction of computer technology, boundary geometries were limited to rather simple patterns, such as straight lines, squares, rectangles, and triangles. The production of circles was a particular challenge. With the introduction of the computer, the options of boundary geometries are limited only by the skill and imagination of the editor. They include the simulation of three-dimensional motion, such as turning the pages in a book.

The technology of generating electronic keying signals developed in parallel with the technology of electronic image generation (see Sec. 8.6). Both have become highly complex, and a complete description is beyond the scope of this volume.

8.5.2   Chroma keying

The preceding section described the electronic production of keying signals to define effects boundaries. It is also possible to create boundary keying signals from images by techniques such as chroma keying.

To illustrate chroma keying, suppose that an image of an individual's face is to be inserted in a landscape. A television image of the face in front of a very bright, highly saturated blue background is recorded. The high-chroma signal from the background area generates a keying signal at the boundary of the signal from the face image. When this keying signal is used in an effects system, the face will appear to be inserted in, not superimposed on, the scene background.

This technique is generally effective, although it requires careful adjustment. In the example above, for instance, if the individual has bright blue eyes, there is a risk that they will generate keying signals, and the edited image of the individual will have portions of the landscape showing through his eyeballs!

8.6   Electronic Image Generation

As with electronic editing, the computer and digital processing have created a major technology for the generation of electronic images. At the present time, the images seldom have a gray scale, and they can be more accurately described as *animation,* but they have found wide use in commercial broadcasting and in the preparation of training tapes.

Also as with electronic editing, the technologies of electronic image generation are complex and often proprietary; see the references. They will not be described here except to note that they are in use and of growing versatility and importance.

Character generators are an important class of image generators. They create signals that produce alphanumeric and other standard characters in a wide variety of fonts and sizes. They are widely used in the broadcast of sports events and the production of commercials.

8.7   SMPTE/ANSI Frame Coding

To create a program by combining material from several sources, it is important that individual frames in each source be identified by a unique code so that they can be quickly located and addressed. The SMPTE/ANSI time code was developed for this purpose. It has two formats, the longitudinal time code (LTC) and the vertical interval time code (VITC). Both have a bit rate of 2400 bits/s and thus can be recorded on an audio track.

Each block in the LTC code has 80 bits: 26 for identifying hours, minutes, seconds, and frame number; 1 for identifying color phase; 1 for identifying dropped frames (108 frames are dropped every hour to convert 30 frame/s to 29.97 frames/s); 32 for user identification; 16 for sync; and 4 undefined.

The VITC has an additional 10 bits per block, for a total of 90. The additional bits are used for a redundancy check and to identify individual fields (rather than frames).

The identification of film frames that are to be integrated with tape recordings poses a number of problems.

The standard time code for film is recorded optically rather than magnetically, but a system has been developed for recording magnetically on the surface of the film.

The basic incompatibility between the 24- and 30-frames/s rates must be accounted for. Computer programs have been developed that automatically make this adjustment so that film and tape frames that are coincident in real time can be brought into coincidence in the editing process.

8.8   Editing Systems

Editing of television programs of any degree of complexity has become a highly skilled and specialized occupation. In addition to his or her purely technical skills, the editor needs a sense of timing and an ability to translate the wishes of the director into the edited output. To carry out this responsibility in an optimum fashion, he or she needs all the technical aids that are available.

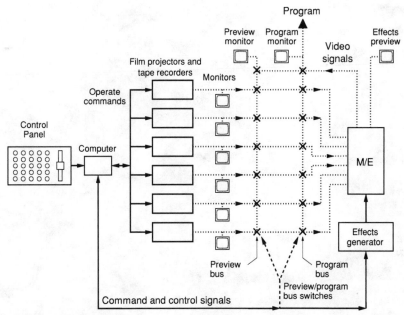

Figure 8.2   Editing system.

Figure 8.2 is a functional drawing that illustrates the elements of a computer-controlled editing system of moderate complexity.

It consists of a number of program sources (usually film projectors and tape recorders), switching and M/E facilities including an effects generator, video monitors for all critical points in the system, one or more keyboard control positions, and a computer. The computer, on command from the editor or from its memory of previous editing decisions, can roll the recorders and projectors (including an allowance for roll time) and control the operation of the switching and M/E facilities. The editor edits the individual program segments that are contained on film and tape, adds effects where appropriate, and assembles the program. Each step is previewed, and the switching and effects commands are recorded in the computer's memory, so that the final product can be the result of repeated fine tuning. An important computer function is to allow for film projector and tape recorder roll times. The computer can also prepare or respond to a hard-copy *edit decision list* that preplans or makes a record of these commands.

Reference

1. K. Blair Benson, (ed.) *Television Engineering Handbook*, Chap. 20, McGraw-Hill, New York, 1986.

Television Transmitting Systems

9.1 Regulatory Framework

Television broadcasting utilizes a portion of the electromagnetic spectrum, and in the United States it is subject to regulation by the Federal Communications Commission (FCC).

On an international level, the United States is a signatory to a treaty, the International Telecommunications Convention, which binds it to follow the rules of the International Telecommunications Union (ITU). The ITU meets infrequently, and between meetings its policies are set by the World Administrative Radio Conference (WARC) and the Regional Administrative Conference (RARC).

All countries have governmental organizations that perform the same regulatory function as the FCC, but in many countries the telecommunications facilities are owned by the government and the government regulates itself.

9.2 Allocations and Assignments

The cognizant regulatory agency in each country *allocates* bands of frequencies to specific services, subject to international agreement. For example, in the United States, the VHF bands 54–72, 76–88, and 174–216 MHz and the UHF band 470–806 MHz are allocated for television broadcasting. (The entire VHF portion of the spectrum extends from 30 to 300 MHz, while the UHF portion extends from 300 to 3000 MHz.) These bands are divided into *channels,* 6 MHz wide, which are *assigned* to specific licensees. Each channel is identified by a number, as shown in Table 9.1.

TABLE 9.1   Television Broadcast Channels

Channel	Frequency, MHz	Channel	Frequency
Low-Band VHF		UHF (cont.)	
2	54–60	35	596–602
3	60–66	36	602–608
4	66–72	37	608–614
5	76–82	38	614–620
6	82–88	39	620–626
High-Band VHF		40	626–632
7	174–180	41	632–638
8	180–186	42	638–644
9	186–192	43	644–650
10	192–198	44	650–656
11	198–204	45	656–662
12	204–210	46	662–668
13	210–216	47	668–674
UHF		48	674–680
14	470–476	49	680–686
15	476–482	50	686–692
16	482–488	51	692–698
17	488–494	52	698–704
18	494–500	53	704–710
19	500–506	54	710–716
20	506–512	55	716–722
21	512–518	56	722–728
22	518–524	57	728–734
23	524–530	58	734–740
24	530–536	59	740–746
25	536–542	60	746–752
26	542–548	61	752–758
27	548–554	62	758–764
28	554–560	63	764–770
29	560–566	64	770–776
30	566–572	65	776–782
31	572–578	66	782–788
32	578–584	67	788–794
33	584–590	68	794–800
34	590–596	69	800–806

The VHF channels are divided into two segments, low-band (channels 2 to 6) and high-band (channels 7 to 13). The small gap between channels 4 and 5 is reserved for aeronautical use.

9.3   The FCC Standard Television Broadcast Channel

The amplitude/frequency characteristic of the 6-MHz television broadcast channel as specified by the FCC is shown in Fig. 9.1.[1] It has two

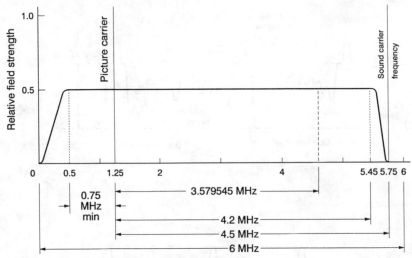

Figure 9.1 Standard U.S. television broadcast channel.

carriers, the visual carrier 1.25 MHz above the lower channel edge and the aural carrier 4.5 MHz above the visual.

9.4 The Visual Carrier

9.4.1 Modulation method and power rating

The visual carrier is amplitude-modulated by the video signal (see Fig. 9.2). The video signal has a dc component, unlike the audio signal in radio broadcasting. It is transmitted by *clamping* (fixing) the carrier power at its maximum rated value, 100 percent, at the peak of sync and setting the gain of the system so that its value at the reference white level is 7.25 percent. See Fig. 9.2.

The *rated* transmitted power is its value at the peak of sync. The *average* transmitted power is less and will depend on the picture content, being least for an all-white picture and greatest for all black. The NTSC video waveform is inefficient in its power utilization, and most of it is used for the sync and blanking pulses; see Sec. 1.8.

9.4.2 Vestigial sideband format

The spectrum of the visual carrier and its sidebands is shown in Fig. 9.1. Both the upper and lower sidebands are transmitted for video signal frequency components up to 0.75 MHz. The lower sideband is attenuated for frequency components above 0.75 MHz, and only the up-

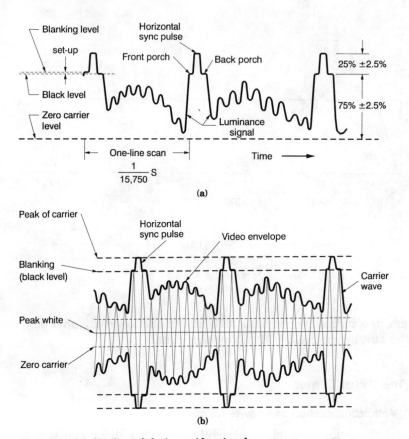

Figure 9.2 Amplitude modulation—video signal.

per sideband is transmitted for signal components above 1.25 MHz.*
This is known as a *vestigial sideband* format, and its purpose is to reduce the amount of spectrum space required for the channel.

The effect of transmitting only a single sideband is shown in the
vector diagrams in Fig. 9.3. Double-sideband operation is shown in
Fig. 9.3*a*. The sidebands are represented by two vectors rotating in opposite directions around the carrier vector. When both sidebands are
present, the amplitude of the vector sum varies by ± twice the
sideband amplitude, and its phase is constant. If one sideband is removed, the amplitude variation of the resultant is reduced by one-half
and phase modulation known as *quadrature phase distortion* is intro-

---

*Low-power UHF stations with less than 1 kW radiated power are allowed to use
double-sideband transmission, except that the lower color subcarrier sideband must be
attenuated by at least 42 dB.

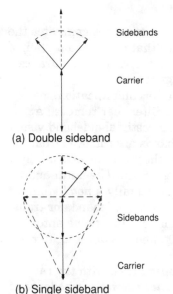

(a) Double sideband

(b) Single sideband

Sidebands

Carrier

Sidebands

Carrier

Figure 9.3 Single-sideband operation.

duced. (Quadrature phase distortion is not the same as *incidental phase distortion*, the variation of the phase of the carrier with level. Incidental phase distortion can be corrected; see Sec. 9.10.3.3.)

The loss in amplitude can be compensated by the amplitude response of the IF amplifier in the receiver (see Sec. 13.4). Typically it is tuned so that the response at the carrier is 6 dB below the maximum. The quadrature distortion cannot be easily corrected, but this is not necessary for the relatively low amplitudes of the high-frequency components in a video waveform. For most picture contents, the maximum high-frequency sideband amplitude occurs for the burst signal, where the phase deviation of the carrier is approximately ±27°.

9.5 The Aural Carrier

The aural carrier is frequency-modulated with a peak deviation of ±25 kHz, as compared with ±75 kHz for FM broadcasting. The lower deviation is specified to make as much spectrum space as possible for the visual signal, and higher deviation is unnecessary because the visual signal establishes the limiting signal-to-noise ratio parameter.

The power of the aural transmitter is normally 20 percent of the peak visual power. This is usually more than adequate to produce an adequate aural S/N in any location with a satisfactory visual S/N.

9.6   Elements of a TV Transmitting System

A TV transmitting system includes a transmitter that generates the modulated visual and aural carriers, a filter that removes a portion of the lower sideband, a diplexer that combines the visual and aural carriers, and an antenna that radiates the signal.

In older systems, the visual carrier, after amplification, passed through a sharply tuned *vestigial sideband* filter that removed a portion of the lower sideband and produced the vestigial sideband visual carrier. In current designs, the lower sidebands are removed from the carrier at an intermediate frequency where the power level is low by a *surface acoustic wave* (SAW) filter; see Sec. 9.8.1.6. The aural and visual carriers are combined in a multiplexer, usually a *notch diplexer* that has high attenuation of the lower subcarrier sideband, for transmission to the antenna in a single transmission line. Some antennas require two feeds separated in phase by 90°, and the phase shifter follows the multiplexer.

In most cases the line is two coaxial copper tubes, with the ratio of their diameters chosen to give the line a *characteristic impedance* of 51.5 or 75 Ω.

The antenna is invariably directional in the vertical plane to concentrate the radiated energy toward the horizon and the service area along the earth rather than wasting it in space.

Some stations, especially in the UHF band, employ antennas that are directional in the horizontal plane. The FCC restrictions on this mode are severe, and antenna directivity cannot be used to permit closer station spacing. It can be used, however, to increase the station's coverage in a desired direction, and this provision of the FCC Rules is often used by UHF stations.

The *gain* of the antenna is the ratio of the maximum radiated energy density to the radiated energy density from a reference antenna with equal power input. The reference antenna used for broadcast systems is the *half-wave dipole*. A half-wave dipole radiating 1 kW in free space produces a field intensity of 137.6 mV/m at a distance of 1 mi.

The half-wave dipole reference differs from the *isotropic radiator*, the reference for microwave and satellite systems. The isotropic radiator is a more fundamental reference, and it simplifies calculations, but it cannot be constructed. The gain of a half-wave dipole is 2.15 dB with respect to an isotropic radiator.

The FCC Rules specify limitations on the *effective radiated power*, or ERP, the visual carrier power input to a half-wave dipole that would result in the same radiated energy density. It is given by the equation

$$\text{ERP} = P_t G_A A_{\text{eff}} \tag{9.1}$$

where ERP = effective radiated power

P_t = output power of the visual transmitter

G_A = gain of the antenna with respect to a half-wave dipole

E_{eff} = efficiency of the transmission lines, diplexer, and other system elements between the output of the transmitter and the input to the antenna

9.7 FCC ERP/Antenna Height Limitations

The FCC limits the ERP and antenna height of television stations to provide the basis for an orderly plan for assigning channels to cities in the United States. These limits are higher in high-band VHF, channels 7 to 13, than in low-band, and higher in UHF than in VHF. This reflects the less favorable propagation characteristics of the higher frequencies. In consideration of these limitations, the FCC was able to establish criteria for mileage separation between co-channel and adjacent-channel stations to minimize interference while permitting a reasonable number of assignments. The radiated power limitations for antenna heights up to indicated maximums are shown in Table 9.2.

TABLE 9.2   FCC Power/Antenna Height Limits*

Channels	Maximum ERP, kW	Maximum height, ft†
2–6	100	1000
7–13	316	2000
14–83	5000	2000

*The power limitations apply to the horizontally polarized component of the radiation. Stations employing circular polarization are permitted to radiate an added equal amount of vertically polarized energy.

†See Sec. 9.15.2 for the definition of antenna height. See Figs. 9.4 and 9.5 for power limitations if antenna height exceeds the maximum.

Additional power limitations apply for heights above these maximums, as shown in Figs. 9.4 and 9.5. The zones on which the limitations are based are defined in Part 73.699 of the FCC Rules. Zone I includes the Northeast and Middle West. Zone III includes the Gulf Coast states, while Zone II includes the remainder of the country.

9.8 Television Transmitters

The designs of television transmitters differ widely depending on their frequency band and power rating. Other differences result from manufacturers' proprietary choices for the basic system arrangement and components. All transmitters, however, must include the following components:

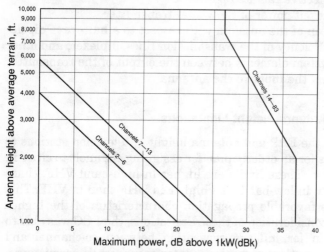

Figure 9.4   FCC power/antenna height maximums (Zone I).

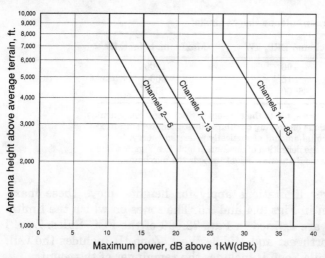

Figure 9.5   FCC power/antenna height maximums (Zones II and III).

1. **Aural preemphasis** to preemphasize the high-frequency components in accordance with FCC requirements.

2. **Processing amplifiers** to correct deficiencies, e.g., improper sync level, in the input visual signal and to provide dc restoration.

3. **Visual and aural carrier generators** with the frequency tolerances specified by the FCC.

4. Visual and aural modulators.

5. Filters to remove the lower sidebands of the modulated visual carriers and to remove spurious frequency components from both carriers.

6. Linearity correction devices in the visual chain to compensate for incidental phase modulation and differential gain. Ideally, these corrections should be made in both the video and RF domains.

7. Intermediate and final power amplifiers.

Figure 9.6 is a functional diagram of a representative television transmitter that shows the locations of these functions.

9.8.1 Exciter modulator

The exciter modulator is the heart of the transmitter and includes functions (1) to (6) in Sec. 9.8. Its outputs are the modulated visual and aural carriers at low power levels.

9.8.1.1 Aural preemphasis. The aural preemphasis is in accordance with FCC requirements. The time constant of the preemphasis circuit is 75 μs.

9.8.1.2 Stabilizing amplifier. Sync pulses are regenerated and amplitude stabilized. The signal is dc restored by clamping the signal at black level. White peaks are clipped to avoid overloading the modulator.

9.8.1.3 Linearity and frequency response correction. Linearity and frequency response correction is made in both the video and RF domains.

9.8.1.4 Aural and visual carrier generation. The aural and visual carriers are generated at intermediate frequencies. Their frequency toler-

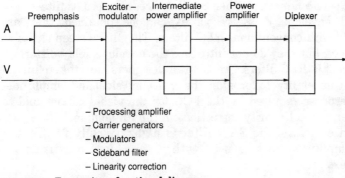

Figure 9.6 Transmitter functional diagram.

TABLE 9.3   Carrier Frequency Tolerances

	Visual	Aural
FCC requirements	±1000 Hz	(Visual + 4.5 MHz) ± 1000 Hz
Industry objective	±200 Hz	(Visual + 4.5 MHz) ± 200 Hz

ances are small (see Table 9.3), in part because of the use of offset carriers (Sec. 2.17.2), which require the difference in carrier frequencies
between two adjacent co-channel stations to be stabilized within ±10
kHz. Stations are assigned frequencies on this basis, and the technique would not be effective if the difference in carrier frequencies
greatly exceeded this tolerance.

It is equally important that the frequency difference between aural
and visual carriers be maintained within close tolerances. This is accomplished by maintaining their frequency difference at 4.5 MHz so
that the carrier frequencies drift together.

9.1.8.5 Visual modulator. The engineering profession was originally
divided on the issue of low-level vs. high-level modulation, i.e., modulating the carrier at a low or a high power level. With the passage of
time, technical advances have favored low-level modulation, and as of
this writing most new transmitters employ this modulation mode.

The advantages of low-level modulation are numerous. They include the elimination of high-power modulators and vestigial
sideband filters, and the ability to correct linearity and frequency response distortions with compensation in both the video and radio frequency domains.

9.8.1.6 SAW filter and sideband removal. The device that finally tipped
the balance for low-level modulation was the *surface acoustic wave*
(SAW) filter.[2] SAW devices are tuned amplifiers or filters that employ
piezoelectric semiconductors. When SAW devices are used as filters, amplitude and envelope delay responses can be obtained that are
unachievable with conventional *RLC* filters. They have steep skirts at
the band edges, but they do not introduce the envelope delay distortion
that is inherent in *RLC* filters. Transmitter designers have taken full advantage of this ability to provide the vestigial sideband amplitude-
frequency response required by the FCC for the visual carrier and to
make frequency and linearity corrections in the RF domain.

The insertion loss of the SAW filter is high, typically 25 dB, and
low-noise amplifiers must be used with it to achieve a satisfactory
signal-to-noise ratio.

9.8.1.7 Up-converter. The up-converter is a conventional heterodyne circuit that mixes the outputs of the visual and aural modulators with that of a local oscillator operating at the sum of the IF and final carrier frequencies. The up-converter outputs at the final carrier frequencies are then sent to the visual and aural intermediate power amplifiers.

9.8.2 Power amplifier drivers

Power amplifier drivers are linear amplifiers that provide the input to the final power amplifier, the PA. They must have sufficient output power to drive the PA to its full rated power, and their frequency response and linearity must be consistent with the performance specifications of the transmitter.

The power requirements vary from less than 25 W for klystron UHF power amplifiers to nearly 2 kW for some tetrode VHF power amplifiers. In common with the trend for all electronic devices, electron tubes are being replaced by solid-state devices as soon as the latter can meet the power level and other performance specifications at an economic price. The technology of solid-state devices has now progressed sufficiently to meet these conditions, and most new commercial transmitters use solid-state drivers.

9.8.3 Power amplifiers

In common with all television technologies, transmitter power amplifiers have benefited from continuous technical improvement. Early VHF transmitters used tetrodes, grid-modulated when high-level modulation was employed. Different models of early UHF transmitters employed both tetrodes and *klystrons,* a tube that employs *velocity modulation* of an electron beam to amplify the carrier. The features of klystrons include stability, high output power, reasonable efficiency (about 40 percent), and high gain, so that very little driving power is required. This characteristic eliminated the need for a long chain of preamplifiers with their linearity problems and was one of the factors that hastened the trend to low-level modulation. Unlike tetrodes, whose performance deteriorates with increasing frequency, klystrons work better at higher frequencies. This led to their early choice as the power amplifier for UHF transmitters.

A hybrid tube, the *klystrode,* has been introduced recently that employs both a control grid and velocity modulation. This has extended the use of the klystron principle into VHF transmitters.

In the meantime, solid-state technology continues to advance, and in time, solid-state devices will probably supersede electron tubes of all types. They are already being used in PA drivers and in low-power PAs, and their power and frequency limits are continuously being expanded.

Solid-state devices used in transmitters are not particularly efficient, and a considerable amount of heat must be dissipated. The *heat pipe* is one solution. A liquid in a wick transfers heat from the device to cooling fins that can be fan cooled if necessary.

9.9   Representative Transmitter Power Ratings

Although the FCC does not specify power ratings for transmitters, certain power ranges have been established for commercial transmitters that are strongly influenced by the maximum ERP permitted by the FCC. Table 9.4 shows typical power ratings for commercial TV trans-

TABLE 9.4   Representative TV Transmitter Power Ratings (kW)

VHF low band	VHF high band	UHF
5	15	30
15	25	60
25	40	120

mitters in the United States. Note that they become progressively higher in the higher-frequency bands. This is not because of equipment limitations, but because the propagation characteristics of higher frequencies are not as favorable—a fact that is recognized in the power limitations specified by the FCC.

9.10   Transmitter Performance Standards

It is important to make a distinction between *equipment* performance and *system* performance in interpreting performance standards. Degradations and distortions are cumulative in analog systems, and for many criteria, e.g., signal-to-noise ratio, the performance of each element in the system must be better than that of the total system. The standards for transmitter performance in this section are for the transmitter only. The standards for the radiated signal, which include the effect of other components of the system, would be somewhat lower. The transmitter standards should be compared with the EIA standards for point-to-point transmission systems.[3]

The following paragraphs summarize the most significant performance standards for television transmitters as contained in two sources: the FCC Rules, Sec. 73, and informal industry equipment *goals* that have been established based on the actual performance of high-quality transmitters.[4] The industry goals are more complete and generally more stringent than the FCC Rules.

9.10.1 Frequency-related standards

9.10.1.1 Carrier frequency tolerance. See Table 9.3 for FCC requirements and industry goals for frequency tolerances.

9.10.1.2 Amplitude-frequency response. The amplitude- frequency response of the transmitter as specified by the FCC is shown in Fig. 9.1. The goals for the maximum deviation from this response (as measured with a sideband analyzer) are shown in Table 9.5.

TABLE 9.5   Amplitude-Frequency Response Tolerance

0.5–4.1 MHz ± 1.0 dB
4.1–4.18 MHz, VHF + 0.0, – 2.0 dB
4.1–4.18 MHz, UHF + 0.0, – 4.0 dB

The upper sidebands between 4.75 and 7.75 MHz and the lower sidebands between 1.25 and 4.25 MHz should be attenuated at least 20 dB below the visual carrier, using the 200-kHz sideband as reference.

9.10.1.3 Envelope delay vs. frequency. The standard *envelope delay* (see Chap. 2) and tolerances are shown in Table 9.6.

9.10.1.4 Frequency response vs. brightness. Measurements of frequency response are made with a modulating sine wave having a peak-to-peak amplitude of 25 percent of reference black to reference

TABLE 9.6   Standard Envelope Delay vs. Frequency

Frequency, MHz	Delay	
0.05–3.0	0 ns	
3.0–4.18	Increases linearly with frequency, with 0 at 3.0 MHz and – 170 ns at 3.58 MHz.	
	Tolerance Goals	
Frequency, MHz	FCC, ns	Goal, ns
0.05–2.1	± 100	± 50
2.1–3.0	± 70	± 35
3.0–3.58	± 50	± 25
3.58–4.18	± 100	± 60

white and with axes on the brightness scale of 25, 45, and 65 percent. The reference is the 45 percent axis.

Tolerance: FCC—not specified; goal—±1 dB.

9.10.2 Signal-to-noise ratio, visual signal

The visual signal-to-noise ratio is defined as S (blanking to reference white)/N (weighted rms) dB.
See Sec. 2.16.3 for the definition of weighted noise.

Tolerance: FCC—not specified; goal—55 dB.

9.10.3 Linearity standards

Linearity standards specify the variations in gain or phase delay with brightness or other signal parameters.

9.10.3.1 Differential gain and low-frequency nonlinearity. Differential gain is the measure of variations in the transmitter's gain at different brightness levels at the color subcarrier frequency; see Sec. 3.17.3. It is measured by superimposing a 3.58-MHz sine wave, having an amplitude of 20 percent of blanking-to-reference white, on a stair-step signal and measuring the differences in its amplitude at the transmitter output on each of the steps.

Low-frequency nonlinearity is the measure of the variation in the transmitter's gain with brightness for lower video frequencies. It is also measured with a stair-step waveform, and the nonlinearity is indicated by variations in the height of the steps at the transmitter's output.

Tolerance: FCC—not specified; goal—5 percent.

9.10.3.2 Differential phase. Differential phase is the complement of differential gain and is the variation in phase delay with brightness; see Sec. 3.17.4. The same waveform is used for measurement.

Tolerance: FCC—not specified; goal—±3°.

9.10.3.3 Incidental phase modulation. Incidental phase modulation differs from differential phase distortion in that differential phase refers to the phase shift of the color *subcarrier envelope,* whereas incidental phase modulation refers to the phase of the *carrier.* It is usually larger than differential phase.

Tolerance: FCC—not specified; goal—±7°.

9.11   Transmission Lines

The output energy of the transmitter is conveyed to the antenna by means of a *transmission line*. This is usually two coaxial copper tubes that are insulated from each other, although *waveguides* are occasionally used for high-power UHF stations. See Sec. 9.12.

9.11.1   Coaxial line transmission mode

When the radio-frequency output of the transmitter is applied to the input of a coaxial line with a gaseous dielectric and a diameter that is small compared with the wavelength, it generates a voltage and current wave that travels at approximately the velocity of radiation in free space. (The deviation results from the fact that the transmission line is not a perfect conductor. In most practical cases, it is small enough to be ignored.) If the dielectric is solid, the velocity and the wavelength are reduced by the square root of the dielectric constant.

In this primary mode, the electric field is perpendicular to the axis of the line. If the wavelength is equal to or less than the circumference of the line, a secondary mode can occur in which the electric field is circular and coaxial with the line. This is known as *multimoding*. Multimoding is undesirable because it increases the losses and reduces the power handling capability of the line. In practice, this limits the usefulness of coaxial lines to frequencies with wavelengths greater than the circumference of the line, a factor that must be considered at the upper end of the UHF band.

The voltage and current waves are in phase, and their ratio, E to I, is Z_c, is the *characteristic impedance* of the line. If the impedance of the load at the end of the line equals the characteristic impedance, all the energy in the wave is absorbed. If the load impedance does not equal the characteristic impedance, some of the energy is not absorbed, and voltage and current waves are reflected back to the transmitter. The magnitude and phase of the voltage and current at each point on the line is the vector sum of the direct and reflected waves, and a stationary *standing wave* is formed, with maxima and minima separated by one-half wavelength.

9.11.2   Voltage and current relationships

Figure 9.7 shows the relationship between the voltage and current vectors at the load and one-quarter wavelength toward the source for four load impedances: the characteristic impedance, a short circuit, an open circuit, and double the characteristic impedance.

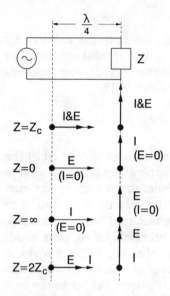

Figure 9.7 Transmission line voltage and current vectors.

The ratio of the voltage maximum to the voltage minimum in the standing wave is called the *voltage standing-wave ratio,* or VSWR. It is a fundamental measurement of transmission line performance.

The VSWR is calculated by Eq. (9.2):

$$\text{VSWR} = \frac{1 + |C_R|}{1 - |C_R|} \qquad (9.2)$$

where $|C_R|$ is the magnitude of the *reflection coefficient.*

C_R is calculated by Eq. (9.3):

$$C_R = \frac{Z_{\text{load}} - Z_c}{Z_{\text{load}} + Z_c} \qquad (9.3)$$

where Z_{load} and Z_c are the load and characteristic impedances. Z_c has a small reactive component as a result of the ohmic resistance of the transmission line conductors; this is usually so small that it can be ignored.

If Z_{load} equals Z_c, the standing-wave pattern disappears, and the voltage and current are constant except for a gradual diminution toward the load owing to losses in the line.

If Z_{load} is a pure resistance and greater than Z_c, a voltage maximum will appear at the load and at 180° intervals toward the load on the line. Similarly, if Z_{load} is a resistance less than Z_c, a voltage minimum will appear at the load and at 180° intervals along the line. If Z_{load} has a reactive component, the maxima and minima will be shifted so that the standing wave at the load will be neither a maximum nor a minimum in the standing-wave pattern.

The input impedance undergoes an inversion every quarter wavelength. For example, an open-circuit quarter-wavelength line will have zero input impedance, and a short-circuited quarter-wavelength line will appear to be an open circuit.

The relationships between load impedance, VSWR, and input impedance in the general case can be determined from a *Smith chart*.

9.11.3 Characteristic impedance calculation

The characteristic impedance of transmission lines is $(L/C)^{1/2}$, where L and C are the inductance and capacitance of the line per unit length. It is determined by the ratio of the diameters of the conductors and the dielectric constant of the insulating material between them:

$$Z_c = \frac{138 \log_{10} (b/a)}{\mu^{1/2}} \tag{9.4}$$

where b and a are the diameters of the outer and inner conductors and μ is the dielectric constant of the insulator between them. The ratio of the outer and inner conductor diameters for lines with an air dielectric, $\mu = 1$, is 3.52 for a Z_c of 75 Ω and 2.30 for a Z_c of 50 Ω.

9.11.4 Standard transmission line diameters

The de facto standard dimensions that have been developed for rigid coaxial transmission lines are tabulated in Table 9.7.

TABLE 9.7   Transmission Line Diameters

Type	Z_c, Ω	Outer diameter, in
560	75	⅞
561	75	1⅝
562A	75	3⅛
563	75	6⅛
573	50	6⅛

Rigid transmission lines normally come in 20-ft lengths and are joined by flanges welded to their ends. A wide variety of elbows, T connections, and other fittings are available.

9.11.5 Power ratings and attenuation

The power ratings and attenuation of transmission lines are normally established for a VSWR of 1.0. If the VSWR exceeds 1.0, the higher

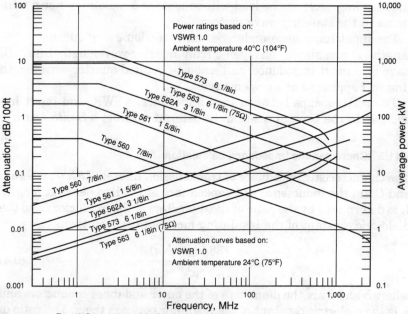

Figure 9.8   Coaxial transmission line power ratings and attenuation.

currents and voltages at their maxima will cause greater attenuation and a lower power rating.

Figure 9.8 shows the power rating and attenuation as a function of frequency for the standard transmission line sizes in the VHF and UHF television frequency ranges, all with a VSWR of 1.0. The rapid increase in attenuation for 6⅛-in line at the upper end of the UHF band is caused by multimoding.

Note that the power ratings are based on *average* transmitter power. The peak-power rating used for television transmissions would be higher, but on the basis of an all-black picture and after allowing for the aural carrier, the difference is small.

9.11.6 Mechanical features

A number of mechanical features of coaxial transmission lines should be noted.

The inner and outer conductors are subjected to different temperatures, so that differential expansion occurs. On a long run the difference can be large, and provision must be made to compensate for it. One solution is to join the inner conductors with slotted plugs that fit inside the two ends. This allows the conductors to expand or contract without breaking the electrical connection.

Provision must also be made for expansion of the outer conductor, and the line is usually hung from the tower with spring hangers rather than being rigidly attached to it.

Moist air inside the line can cause condensation and leakage across the inner conductor spacers. To prevent this, most lines are pressurized with dry air. This requires that the joints be sealed so that they are air tight.

9.12 Waveguides

It is unfortunate that the trends of the power capacity and attenuation vs. frequency curves of coaxial transmission line are in the wrong direction for UHF transmissions. More power is required for UHF channels, but the power rating of coaxial lines is lower at the higher frequencies, and their attenuation is greater. The use of larger-diameter line at the upper end of the UHF band to reduce losses and increase the power rating is limited by multimoding. As a result, waveguides, which are better suited for higher frequencies, are sometimes used for the upper UHF channels.

Rectangular waveguides operating in the TE_{10} mode (see Fig. 9.9) are the most commonly used. In this mode, the width of the waveguide must be more than one-half wavelength at the transmission frequency. The frequency at which the wavelength equals twice the width of the waveguide is its *cutoff frequency,* and the waveguide will not transmit lower frequencies.

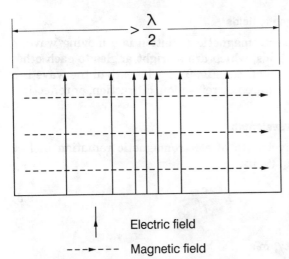

    Electric field

- - - ► - - -   Magnetic field

Figure 9.9 TE_{10} waveguide transmission mode.

TABLE 9.8   Waveguide Power Rating and Attenuation, TE_{10} Mode

Type	Width inside, in	Frequency range, MHz	Power rating, MW	Attenuation, dB/100
WR 1800	18.0–9.0	410–625	93.4–131.9	0.056–0.038
WR 1500	15.0–7.0	490–750	67.6–93.3	0.069–0.050
WR 1150	11.5–5.75	640–960	35.0–53.8	0.125–0.075

The losses in waveguides are much lower than those in transmission lines at these frequencies. The dimensions, frequency range, power rating, and attenuation of aluminum waveguides operating in the TE_{10} mode and in the standard sizes that cover the UHF broadcast band are shown in Table 9.8. Note that the attenuation *decreases* and the power rating *increases* as the frequency increases.

Like coaxial transmission lines, waveguides are subject to condensation of moisture as they cool. This can be avoided by pressurization, but the amount of pressure is limited by the mechanical weakness of the rectangular structure. Higher pressures can be used with *double truncated waveguide*. The corners of a rectangular waveguide are rounded, and it is fitted inside a circular tube. The circular structure not only permits greater pressurization, but also reduces the wind loading.

9.13   Properties of VHF and UHF Radiation

This section provides a brief review of the properties of VHF and UHF electromagnetic radiation as an introduction to a description of television broadcast antennas and the coverage areas of broadcast stations.

9.13.1   Electric and magnetic fields

As its name implies, electromagnetic radiation is a moving wave of electric and magnetic fields, which are at right angles to each other and to the direction of travel. See Fig. 9.10. The path of the wave is a straight line unless it is altered by refraction, diffraction, or reflection.

9.13.2   Frequency and wavelength

The frequency and wavelength of electromagnetic radiation in free space are related by Eq. (9.5):

$$\lambda = \frac{c}{f} \tag{9.5}$$

where λ = wavelength, m
    c = wave velocity, m/s
    f = frequency, Hz

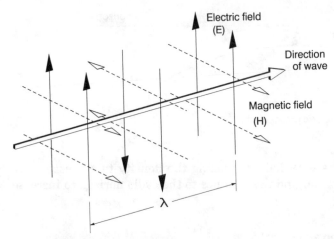

Figure 9.10 Electromagnetic wave.

9.13.3 Velocity and index of refraction

The velocity of electromagnetic radiation in free space, c, is 3×10^8 m/s. Its velocity through gases, liquids, or solids is c/n, where n is the *index of refraction* of the transmission medium. The index of refraction of the atmosphere varies in accordance with altitude, barometric pressure, and humidity. Typically it is 1.003 at sea level and decreases exponentially with altitude. The deviation from unity is small, but is sufficient to have a major effect, normally an increase, on the coverage of television broadcast stations because of *refraction*.

9.13.4 Refraction

Refraction occurs when the radiation path passes obliquely across a boundary between two media that have different refractive indices. The path is bent toward the denser medium, which has a higher index of refraction.

The path of a broadcast television signal is directed tangentially along the earth's surface at the transmitter and passes through progressively less dense layers of the *troposphere,* the lower portion of the atmosphere. Because of refraction, it is bent downward and passes beyond the geometric horizon; this is known as tropospheric transmission or "tropo" (see Fig. 9.11). Under normal atmospheric conditions, the effect of refraction can be approximated by assuming that the earth's radius is ⁴/₃ of its actual value, thus increasing the distance to the radio horizon. Atmospheric abnormalities sometimes exist in which the atmospheric density *gradient* deviates from its normal pattern. The most common is an *inversion,* in which the temperature of

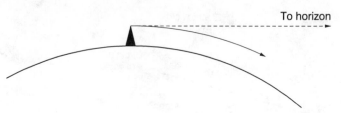

Figure 9.11   Tropospheric propagation.

the air increases with height, causing the density to decrease more rapidly than normal and the distance to the radio horizon to increase correspondingly.

9.13.5 Diffraction and reflection

Electromagnetic radiation travels in a straight line in free space, but if part of the wavefront is blocked by an intervening object, the path will bend, and some radiant energy will appear behind the geometric shadow of the object. This is one of the effects of a phenomenon known as *diffraction*.

Radiant energy is reflected, and, as the result of the combined effects of reflection and diffraction, television signals can be received at most points in a station's service area even though few of them have a line-of-sight path to the transmitting antenna.

9.13.6 Multipath transmission

The signal at the receiving antenna is the vector sum of a direct signal and one or more reflected signals. If the difference in path lengths is small, the signal strength will be increased or decreased depending on the relative phase of the incoming signals. If the difference in path lengths is large, e.g., as the result of a reflected signal from a large, distant object, the time difference will be great enough to cause a *ghost*. This effect is called *multipath,* and it can be serious in mountainous or congested urban areas.

9.13.7 Polarization

The *polarization* of a wave describes the orientation of its field vectors, and by convention it is the orientation of the electric field. Two polarization modes are permitted for television broadcasting, horizontal and elliptical. In elliptical polarization, the electrical vector rotates as the wave progresses and passes a fixed point. Circular polarization is a special case of elliptical in which the vector has an equal value in all orientations.

FCC Rules permit the same ERP for the horizontal component of circularly polarized radiation (the component on the major axis for elliptical polarization) as for horizontal polarization, and the total radiated power can be twice as great.

Circular polarization has a number of advantages. Since the total radiated power is twice as great, the received signal is usually greater with "rabbit ears" or other indoor antennas that are not optimized for horizontal polarization. Improvements of up to 3 dB in received signal strength can be achieved by using circularly polarized receiving antennas that derive energy from both the horizontal and vertical components. Circularly polarized receiving antennas also reduce the magnitude of ghosts, since the direction of rotation is inverted upon reflection, and the reflected wave is not received. As a result of these advantages, there has been an increasing trend toward the use of circular polarization.

9.13.8 Radiated field strength and energy

The intensity of electromagnetic radiation is usually specified by the strength of its electric field. The usual units for the magnitude of the fields encountered in broadcasting are millivolts per meter or microvolts per meter.

The field strength of an electromagnetic wave in free space is inversely proportional to the distance from the antenna. For radiation from a half dipole, it is

$$E = \frac{137P^{1/2}}{D} \tag{9.6}$$

where E = field strength, mV/m
P = radiated power, kW
D = distance, mi

The strength of a wave can also be measured by the power passing through a unit area, usually a square meter, at right angles to the wave's path. Field strength and energy in space or in the atmosphere are related by Eq. (9.7):

$$\text{Power/m}^2 = E^2/377 \tag{9.7}$$

where power is measured in microwatts and E is the rms value of the field strength measured in millivolts per meter.

The constant 377 (120 π) has the dimension of ohms, and it is sometimes described as the impedance of space.

9.14   Antennas

The role of the antenna is to couple the transmitter to space by converting its electrical output energy to electromagnetic radiation. The primary antenna specifications are vertical directivity and gain, horizontal directivity, polarization, input impedance and bandwidth, and power handling capacity.

9.14.1   Vertical directivity and gain

Television transmitting antennas increase the signal strength on the earth's surface by minimizing the power that is radiated into space and concentrating it at angles toward or just below the horizon. This property, the antenna's *directivity*, is indicated by a *radiation pattern*, a plot of the radiation intensity at different elevations.

Figure 9.12 shows the vertical radiation patterns for three representative antennas, a low-band, low-gain VHF antenna (see Sec. 9.6 for the definition of gain), a high-band, medium-gain VHF antenna, and a high-gain UHF antenna.

The points on the pattern where the field strength is 70 percent of the maximum are known as the *half-power points,* and the angular separation of these points is the *beam width* (see Fig. 9.12).

The gain of the antenna is increased by narrowing the beam width to concentrate the radiated power in the desired directions, and the product of the gain and beam width tends to be constant.

The directivity of a television broadcast antenna can be calculated by defining its *aperture,* an imaginary cylinder in space surrounding the radiating elements. The cylindrical aperture is *illuminated* by the radiating elements, and for analytic purposes each point on the aperture can be considered to be a source of radiant energy.

The antenna's directivity is determined by the dimensions of the aperture and by the amplitude and phase of the radiant energy at each

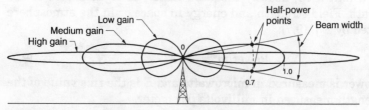

Figure 9.12   Vertical antenna radiation patterns.

point on its surface. In the simplest case, all the points on the aperture are illuminated by a field of the same phase and of equal amplitude. The beam width, the angular separation between the half-power points at which the radiation intensity is 70 percent of that at the beam maximum, is inversely proportional to the ratio of the aperture height to the wavelength. If the illumination of the aperture is uniform, the beam width is

$$\text{Beam width} = 48\left(\frac{\lambda}{h}\right)^{\circ} \tag{9.8}$$

where λ is the wavelength and h is the height of the aperture measured in the same units as λ.

If the illumination of the aperture is not uniform, Eq. (9.8) does not apply precisely, but in general, the longer the aperture relative to the wavelength, i.e., the taller the antenna, the narrower the beam.

9.14.2 Horizontal directivity

The FCC Rules make a distinction between antennas that are intentionally designed to be directional and those that are directional by virtue of unavoidable imperfections in the design. There are no firm rules with respect to the circularity of the horizontal pattern of "nondirectional" antennas, but most of them maintain it within ±3 dB or less.

As the result of its unfortunate experience with AM directional antennas, which were overused to increase the number of stations on each frequency, the FCC ruled that directional antennas *cannot* be used in television to reduce the spacing between stations. Also, the maximum ERP from directional antennas cannot exceed the maximum allowed for nondirectional antennas. The major purpose of directional antennas is to avoid wasting energy over sparsely populated areas and to provide the maximum allowed ERP to densely populated areas with lower total power.

9.14.3 Bandwidth, VSWR, and echoes

Most antenna components are frequency-sensitive, and achieving the proper input impedance over the bandwidth of a single channel was one of the earliest challenges for television technology. In subsequent years, techniques have been developed that make it possible to achieve satisfactory response over more than one channel.

The conventional indicator of antenna bandwidth is the range of frequencies at which its input impedance approximates the characteristic impedance of the transmission line. This is measured by the VSWR of the transmission line feeding the antenna, and it is desirable that this be less than 1.05 for frequencies from 0.75 MHz below to 5.45 MHz above the visual carrier. The VSWR at the color subcarrier 3.58 MHz above the carrier is particularly important.

A low VSWR is not an absolute guarantee of a satisfactory match between antenna and transmission line, and reflections from the antenna may produce visible echoes or ghosts, even though the VSWR is within acceptable limits. The visibility of the echoes depends on their amplitude and their separation from the primary signal. For a transmission line length of 1000 ft, echoes having an amplitudes of 1.5 percent or more will be visible. For a 500-ft length, visibility begins for echoes having an amplitude of 3.0 percent or greater.*

9.14.4 Radiating systems

Antennas can be classified by the type of radiators that illuminate the aperture, and these can be divided into two categories, dipoles and slots.

9.14.4.1 Dipole radiators. This section describes representative dipole radiators.

1. Superturnstile antenna. The *superturnstile* antenna was one of the first to be introduced in the early years of television broadcasting, and it is still in wide use for VHF stations. The dipole is a pair of radiators, sometimes called *bat wings* because of their shape, mounted on opposite sides of a supporting pole. The *radiator* consists of two dipoles mounted at right angles, and the *antenna* consists of two to twelve layers of dipoles. See Fig. 9.13. The spacing between layers is approximately one wavelength, and the gain is approximately equal to the number of layers. The dipoles are plane surfaces rather than cylinders to increase their bandwidth. The dipole pairs are fed in quadrature and generate a reasonably circular pattern in the horizontal plane.

2. Panel antenna. The superturnstile antenna requires a cylindrical mounting pole of limited diameter, and so it is not well suited for the lower element of a stacked array where two or more antennas are mounted above one another. The panel antenna (see Fig. 9.14) solves this problem by mounting the dipole in front of a reflector and mounting a dipole/reflector assembly on each face of a tower that can be either four- or three-sided. Panel antennas are used for both high- and low-band VHF stations.

---

*See Benson, Chap. 8, for a more complete description of antenna echoes.

Figure 9.13   Superturnstile antenna.

3. **Circularly polarized antennas.** In one form of circularly polarized antenna, the single horizontally polarized dipole in the panel antenna is replaced by crossed dipoles that are fed in quadrature, thus producing a circularly polarized wave.

9.14.4.2 Slot radiators. Slot radiators are most frequently used for UHF stations but are also used for VHF. They consist of a hollow cylindrical column in which vertical slots are cut. A conductor in the center of the column, which is connected to the transmitter output, is capacitively coupled to the slots by radial pins mounted on their edges; see Fig. 9.15.

An electric field is produced across the slots, with the amplitude and phase determined by the length and position of the pins. The electric field across the slots and the current around the cylinder generate the radiated energy.

There are two types of slot antennas, *standing wave* and *traveling wave*. The inner conductor of the standing-wave antenna is connected to the outer cylinder at the end opposite the feed point, and a standing-wave pattern is produced. The inner conductor of the traveling-wave antenna is terminated by a capacitive connection to the end slots that absorbs any remaining energy and provides a characteristic impedance load. The amplitude of the voltage in the standing-wave antenna is a standing-wave pattern with diminishing

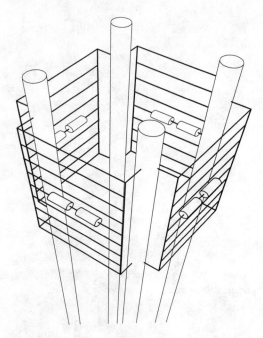

Figure 9.14 Panel antenna.

Figure 9.15 Slot antenna.

maximum amplitudes. The amplitude of the voltage in the traveling-wave antenna diminishes continuously from feed point to termination.

9.14.4.3 Helical radiators. Helical antennas are occasionally used. The radiating elements are two helixes, one above the other, wrapped around but separated from a supporting pole. The two are wrapped in opposite directions, with the result that the vertical components of the radiation from the two halves cancel each other while the horizontal components are reinforced.

9.14.5 High-gain UHF antennas

The gain of VHF antennas is lower than that of UHF, first because the longer wavelengths would lead to impractical heights, and second because the lower power limits for VHF stations reduce the necessity. A gain of 12 is common for high-band VHF antennas. By contrast, a gain of 50 is typical of UHF antennas. Their gain and the accompanying narrow radiation pattern create a number of problems.

High-gain antennas must be adjusted with a high degree of precision because of the leverage involved in the geometry of the radiation path.

Practical antenna radiators can only approximate the theoretical illumination patterns, and the antenna patterns must be verified before the antenna is erected.

If the pattern is adjusted so that its main lobe is horizontal, the pattern maximum will be above the surface of the earth and will miss it, even at great distances. This problem can be solved or alleviated by *beam tilt.*

Nearby areas will receive a weak signal, since they will be in the null of the pattern. This problem can be solved by *null fill.*

9.14.5.1 Beam tilt. The beam is aimed slightly below the horizontal so that it is aimed toward the horizon or slightly below it. The depression angles and distances to the horizon for different antenna heights are:

Height, ft	Angle, degrees	Distance, mi
500	−0.34	31.6
1000	−0.49	44.7
2000	−0.69	63.2

The beam is often pointed below the horizon, and beam tilts of 1.0° or more are common. Figure 9.16 shows a high-gain pattern with 2° of beam tilt. This pattern is based on uniform illumination of the aperture, but the phase varies linearly from +75° at the top of the aperture to −75° at the bottom.

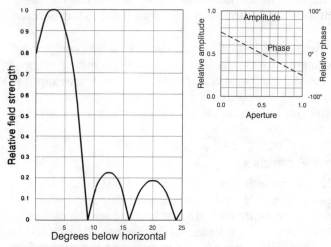

Figure 9.16 Vertical pattern with beam tilt. (*From K. Blair Benson,* Television Engineering Handbook, *McGraw-Hill, New York, 1986, Fig. 8-21.*)

9.14.5.2 Null fill. The second problem is solved by *null fill,* deliberately increasing the signal in the nulls of the pattern to increase its ERP in the angles for close-in points. By careful adjustment, the pattern can be shaped so that the product of distance and ERP$^{1/2}$, the indicator of received field strength, remains fairly constant out to a considerable distance. Figure 9.17 shows the radiation pattern of an antenna with null fill. Here the phase is constant, but the relative amplitude rises exponentially from 0.15 at the bottom to 1.0 at the top.

9.14.6 Power ratings

The power handling limitations of television antennas are established by corona and voltage breakdown and by heating due to its ohmic resistance. Both must be taken into account in the design and construction of the antenna, and the manufacturer's published specifications should be carefully followed.

9.15 TV Station Service Areas

9.15.1 FCC definitions

A TV station's service area includes the locations, usually contiguous, in which the signal level is adequate to produce a satisfactory signal-to-noise ratio in the receiver, and that are not subject to an unsatisfactory level of co-channel or adjacent channel interference. The FCC

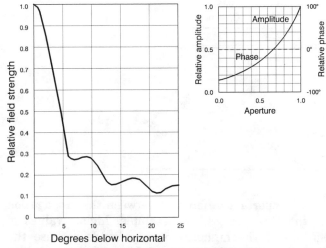

Figure 9.17 Vertical pattern with null fill. (*From K. Blair Benson,* Television Engineering Handbook, *McGraw-Hill, New York, 1986, Fig. 8.23.*)

has established standards for both parameters, and they are generally accepted by the industry.

The specification of the signal level required for a service area must recognize that signal levels vary widely and randomly within relatively short distances because of differences in absorption, reflection, diffraction, and refraction at adjacent points on the highly irregular surface of the earth. At greater distances they also vary with time because the refractive index of the atmosphere changes with the weather.

As the result of these variations, it is necessary to define service areas on a statistical basis, and signal levels are usually defined by the term $F(50,50)$. A level of $F(50,50)$ means that the field strength F is exceeded at 50 percent of the potential receiver locations for 50 percent of the time within a limited area. A receiving antenna height of 30 ft is assumed. F is usually defined in dB_μ, the number of decibels above a field strength of 1 $\mu V/m$.

The FCC defines three grades of television service: Principal Community, Grade A, and Grade B. These grades are defined in Table 9.9. As shown in the table, a different probability of service is defined for each of these grades, but all are specified in terms of the corresponding $F(50,50)$ values.

The specified values are based on the signal levels that were required to overcome both thermal and man-made noise with the monochrome receivers available in the early 1950s.[5] See Chap. 13 for a discussion of recent developments in receiver design.

TABLE 9.9   Grades of Television Service, F(50,50), dB$_u$

Service	Channels		
	2–6	7–13	14–83
Principal community 90% of locations 90% of time	74	77	80
Grade A 70% of locations 90% of time	68	71	74
Grade B 50% of locations 90% of time	47	56	64

Table 9.9 indicates important differences between the propagation characteristics of the three bands. For example, higher values of F(50,50) are required at the higher-frequency bands because the smaller receiving antennas require greater field strength to produce the same voltage at their terminals. Note that the difference between the F(50,50) value for the Principal Community and Grade B services is 27 dB for channels 2 to 6, but only 16 dB at UHF. In part this is because the level of man-made noise is higher in the low-band VHF channels than in UHF, particularly in urban areas.

9.15.2   FCC field intensity curves

The FCC developed and published[1] families of curves on the basis of measurement data that show the values of F(50,50) for a range of distances and transmitting antenna heights. One set of curves is for high-band VHF, channels 7 to 13, and the other, somewhat paradoxically, for low-band VHF and UHF.

The FCC also published F(50,10) curves that show the values of field strength that are occasionally exceeded. These are mainly of value in calculating interference between stations, since a signal level that exists only 10 percent of the time has little value in providing service, but can be very annoying as interference. These curves were used in constructing the FCC Table of Allocations.

It must be emphasized that these curves show statistical results only and do not make allowance for specific terrain features. They have little value, therefore, in forecasting the field strength at a particular point at a particular time. With an understanding of this limitation, however, they fill a useful role in estimating the gross features of television coverage.

Figures 9.18 and 9.19 show the FCC F(50,50) curves for high-band VHF and for low-band VHF and UHF. The abscissa is the antenna

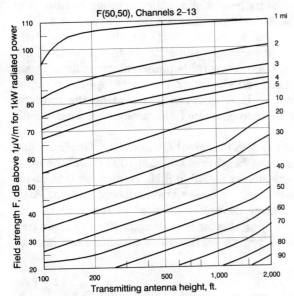

Figure 9.18 FCC F(50,50) field intensity curves, high-band VHF.

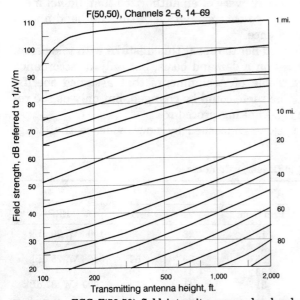

Figure 9.19 FCC F(50,50) field intensity curves, low-band VHF and UHF.

height, which is defined as its height above the "average terrain," the average elevation in the 2- to 10-mi distance from the antenna.

The antenna height has a major effect on the received field strength, in part because it increases the distance to the horizon, but also because it results in a more favorable phase relationship between the direct waves and those reflected from the earth. As an example, according to the curves for UHF coverage, increasing the antenna height from 500 to 1000 ft increases the field strength by nearly 8 dB or 8 times.

The predicted distance to the Principal Community, Grade A, and Grade B contours for antenna heights of 500 and 1000 ft and the maximum permitted ERP are shown in Table 9.10. Achieving this coverage for UHF stations requires a costly transmitting system. Also, UHF transmissions are more adversely affected in rough terrain or built-up areas because of their shorter wavelength. For the same reason, variations in signal strength within short distances are greater and more rapid.

9.16 FCC Table of Assignments

AM radio channels were assigned on the basis of demand. The FCC established co-channel and adjacent-channel interference criteria, and if the applicant for a new station could show that his proposal would meet these criteria, a construction permit and ultimately a license could be granted. But stations were often authorized that did *not* meet these criteria, and this led to overpopulation of the AM band and to excessive mutual interference.

The FCC was determined not to repeat this mistake with television. Stations were authorized on a demand basis, but only in accordance with a predetermined Table of Assignments that assigned specific channels to each city. The table could be changed, but only as the result of a cumbersome rule-making procedure.

In 1949, the FCC became concerned by new evidence that long-

TABLE 9.10   Distance (mi) to Coverage Contours

Channels	Height, ft	Service Grade			ERP
		Prin.Com.	A	B	
2–6	500	19	28	43	100 kW, 20 dBk
	1000	28	37	57	
7–13	500	25	36	52	316 kW, 25 dBk
	1000	33	41	60	
14–83	500	31	41	60	5000 kW, 37 dBk
	1000	41	52	70	

distance tropospheric transmission (see Sec. 9.13.4) was more common than had been previously assumed, and that this would lead to a greater degree of interference between distant stations. It declared a hiatus on granting new licenses, the "freeze," pending further study of tropospheric interference. The freeze ended in 1951 with the issue of a new Table of Assignments that was based on increased spacing between co-channel and adjacent-channel stations, as shown in Table 9.11. This resulted in a major reduction in the number of VHF stations, particularly in the larger cities.

TABLE 9.11   Minimum Co-Channel Station Separation (mi)

Zone	Channels 2 to 13	Channels 14 to 69
I	170	155
II	190	175
III	220	205

Zone I encompasses the northeast tier of states and parts of the middle west. Closer spacing is permitted there in recognition of the closer spacing of cities.

Zone III includes the Gulf Coast states. Greater spacing is required because of the prevalence of inversions in the troposphere and greater tropospheric propagation..

Zone II covers the remainder of the United States.

The table also includes additional criteria or "taboos" for UHF channels based on the characteristics of superheterodyne receivers. For example, no channel is assigned on the image frequency of another channel in the same city.

The present Table of Assignments has generally been successful in minimizing co-channel and adjacent-channel interference (although it may be a problem for HDTV systems).

References

1. FCC Rules, Sec 73.699.
2. Donald G. Fink and Donald Christiansen, Electronic Engineers' Handbook, 3d ed., McGraw-Hill, New York, 1989, Chap. 13.
3. Electronic Industries Association EIA/TIA Standard, Electrical Performance for Television Transmission Systems, EIA/TIA-250-C (1990).
4. K. Blair Benson, Television Engineering Handbook, McGraw-Hill, New York, 1986.
5. Federal Communications Commission, "Sixth Report and Order," Docket Nos. 8736, 8975, 8976, 9175, April 11, 1952.

CATV Systems

10.1 A Brief History of CATV

Cable television or *CATV* began in the very late 1940s as *community TV,* a system for bringing television service to small communities that did not receive adequate off-the-air signals in individual homes, either as the result of their distance from broadcasting stations or because of intervening hills or other terrain features. Local TV dealers or other entrepreneurs erected high-gain antennas on mountain tops or towers to provide usable signals to cable systems that delivered them to local homes.

Within a few years the variety of program services available to community television systems was increased by importing signals from distant stations by microwave, and service was delivered to communities far outside the normal service areas of the originating stations.

In the late 1970s, communication satellites made it possible for cable systems to receive programs from all over the world. The variety of available programming became great enough to be attractive to viewers in large cities, even though they had several off-the-air sources available. At this time, community TV was no longer descriptive of the service, and cable TV or CATV became the accepted nomenclature. CATV is now a large business, and more than 40 million homes in the United States are receiving programs by cable.

10.2 Cable System Elements

Figure 10.1 shows the elements of a cable TV system. The incoming signals—off-air, microwave, and satellite—are received at the *head end.* The carrier frequencies of the off-air signals are usually shifted and multiplexed with the signals that are received from microwave and satellite circuits. The latter are in the baseband mode, and they modulate carriers in accordance with broadcast standards.

The cable distribution system is a combination of *trunks, feeders,*

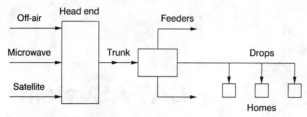

Figure 10.1   Cable system elements.

and *drops*. Trunks are cables that feed major portions of the service areas. They are connected to feeders that serve individual streets and the drops that make the connections to individual homes. At the present time, with the exception of drops, which may employ flat twin lead, these system components employ coaxial cable.

Some systems have *super trunks* to interconnect more distant parts of the system or even different systems. Super trunks may employ co-axial cable, fiber optic cable (see Chap. 12), or a type of microwave known as AML, for amplitude-modulated link.

10.3 Cable Television Channels

The first community antenna systems employed the five low-band VHF channels, 2 to 6, because of their low attenuation (see Sec. 10.4.2.2). As the demand for channels grew, systems expanded into the high-band UHF, channels 7 to 13. The spacing between line amplifiers had to be reduced, but unmodified standard receivers could be used.

As the demand for channels grew still further, the industry had the choice of adding channels in the UHF spectrum or of utilizing channels between low-band and high-band VHF and between high-band and UHF. The latter course was chosen because of the extremely high attenuation in the UHF band.

A standard configuration for CATV provides 55 channels, with the highest channel just below the UHF band. More channels can be added by expansion into the UHF band; see Table 10.1.

Adjacent channels are not intended to be received at the same location in broadcast practice because receivers are unable to discriminate between a weak desired signal and a strong undesired one on an adjacent channel. In CATV systems, however, adjacent-channel signals can be maintained at approximately equal levels and adjacent-channel reception is possible. It requires reasonably careful control of signal levels to avoid imbalances. Also, it requires good control of system linearity to avoid cross-modulation between channels.

Most of the channels in this configuration cannot be received directly by a standard TV set, and *set-top* converters that heterodyne all incom-

TABLE 10.1   CATV Channels

Channel designation	Frequency, visual carrier, MHz	Channel designation	Frequency, visual carrier, MHz
Low-Band VHF		Superband (cont.)	
2	55.25	N	241.25
3	61.25	O	247.25
4	67.25	P	253.25
5	77.25	Q	259.25
6	83.25	R	265.25
MidBand VHF		S	271.25
A-2	109.25	T	277.25
A-1	115.25	U	283.25
A	121.25	V	289.25
B	127.25	W	295.25
C	133.25	Hyperband	
D	139.25	AA	301.25
E	145.25	BB	307.25
F	151.25	CC	313.25
G	157.25	DD	319.25
H	163.25	EE	325.25
I	169.25	FF	331.25
High-Band VHF		GG	337.25
7	175.25	HH	343.25
8	181.25	II	349.25
9	187.25	JJ	355.25
10	193.25	KK	361.25
11	199.25	LL	367.25
12	205.25	MM	373.25
13	211.25	NN	379.25
Superband		OO	385.25
J	217.25	PP	391.25
K	223.25	QQ	397.25
L	229.25	RR	403.25
M	235.25		

ing channels to a single output channel, usually channel 3 or 4, or *cable-ready* receivers that can tune to the cable channels must be used.

10.4   Coaxial Cable

Coaxial cable is almost universally used for the distribution of CATV programs.

10.4.1   Physical construction

The physical construction of a coaxial cable of the type commonly used in CATV systems is shown in Fig. 10.2. The inner conductor is copper-clad aluminum. Copper, with its low resistivity, is used on the outside

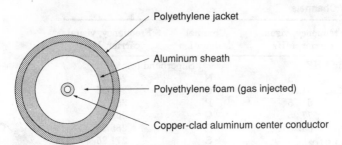

Figure 10.2   Coaxial cable construction. (*From Eugene R. Bartlett*, Cable Television Technology and Operation, *McGraw-Hill, New York, 1990.*)

of the inner conductor where the current density is highest. Aluminum is used for the outer conductor and the core of the inner conductor because of its light weight.

The insulating material between conductors is polyethylene foam. Gas is sometimes injected to prevent water from migrating through it. A flexible steel tubing is sometimes placed over the polyethylene jacket for mechanical protection.

Standard cable diameters are 0.50, 0.75, 0.875, and 1.00 in.

Cables, particularly in the larger diameters, are only semiflexible, and the bending radius is an important specification.

Cable expands and contracts with temperature change, approximately 13 parts per million per degree Fahrenheit, and provision must be made to allow for this in the installation of the cable. The problem is complicated by the fact that the steel messenger that supports an overhead cable expands at a different rate, and allowance must be made for this also.

10.4.2   Electrical characteristics

10.4.2.1   Characteristic impedance.   The characteristic impedance of flexible transmission line has the same electrical significance as for the solid-tube lines used to connect transmitters to antennas (see Sec. 9.6). The standard characteristic impedance for CATV transmission line is 75 Ω.

10.4.2.2   Attenuation.   The signal is attenuated as it passes down the cable owing to ohmic losses. The amount of loss depends on the diameter of the cable, the frequency, the standing-wave ratio, and the temperature.

Table 10.2 shows the attenuation in decibels per 100 feet of cables of the type illustrated in Fig. 10.2 with four different diameters. The attenuations in this table are based on a VSWR of 1.0 and a tempera-

TABLE 10.2 Cable Attenuation (dB/100 ft)

Frequency, MHz	Cable diameter, in			
	0.500	0.750	0.875	1.000
5	0.16	0.11	0.09	0.09
30	0.40	0.26	0.24	0.23
50	0.52	0.35	0.32	0.30
110	0.75	0.52	0.47	0.44
150	0.90	0.62	0.55	0.52
180	1.00	0.68	0.60	0.59
250	1.20	0.81	0.72	0.68
300	1.31	0.90	0.79	0.75
400	1.53	1.05	0.91	0.87
450	1.63	1.12	0.98	0.92
500	1.70	1.17	1.04	0.97
550	1.80	1.23	1.09	1.02
600	1.87	1.29	1.15	1.06

ture of 68°F. Note that the attenuation (in decibels) is approximately proportional to the square root of the frequency.

The attenuation increases with temperature, at the approximate rate of 1 percent per 10°F change.

10.4.2.3 Return loss. The return loss is a measure of the deviation of the actual value of the cable's characteristic impedance from its nominal value. It is equal to the ratio of the incident power P_i to the reflected power P_R, in decibels:

$$L_R \text{ (dB)} = 10 \log \left(\frac{P_i}{P_R} \right) \quad \text{dB} \tag{10.1}$$

The closer the actual characteristic impedance of the line matches its nominal value, the smaller the P_R and the larger the return loss. For a perfect match, $P_R = 0$ and $L_R = \infty$. No line is perfect, and practical values of L_R vary from 28 to 32 dB. If the return loss is too low, reflections occur along the line, causing variations in voltage or *signal ripple.*

10.4.2.4 Loop resistance. The primary power for the line amplifiers that are installed to compensate for its attenuation is usually supplied by low-voltage alternating or direct current transmitted by the cable along with the RF signal. Since the amplifiers operate at relatively low voltage, typically 45 V, and high current, the *loop resistance,* the combined resistance of the inner and outer conductors to direct or alternating current is an important cable specification.

Unlike high-frequency RF signal currents that are conducted along the outer surface of the inner conductor because of the "skin effect," dc and ac power currents are conducted by the entire cross section. As a result, the resistance of copper-clad aluminum conductors is significantly higher than solid copper conductor for power currents. For example, the loop resistance of 0.50-in solid copper cable is 1.23 Ω/1000 ft, whereas it is 1.68 Ω/1000 ft for copper-clad cable of the same diameter.

10.5 Line and Bridging Amplifiers and Equalizers

Line amplifiers are inserted in the cable at intervals to compensate for its attenuation, and their specifications are a key element in determining the performance of the system. Each amplifier includes an equalizer that compensates for the increase in cable attenuation with frequency.

Bridging amplifiers have a high-impedance input and are used to tap the signal from a trunk without disturbing its performance.

10.5.1 Technical requirements

The technical requirements for amplifier/equalizer combinations are demanding and are even more stringent because of the cumulative degradation of a large number of amplifiers in series.

1. They must operate over a wide frequency range, and their gain must be high enough at the top of the frequency range to enable the amplifiers to be separated by a reasonable distance.

2. The equalizers must compensate for the variations in cable attenuation with frequency with a considerable degree of precision.

3. The linearity of the amplifiers must be high to minimize crosstalk and the generation of beat frequencies between channels.

4. Automatic gain and frequency response, *slope correction,* must be provided to compensate for variations in these parameters with temperature or equipment aging; see Sec. 10.5.2.

5. The signal-to-noise ratio of individual amplifiers must be high enough so that the cascaded noise level of all the amplifiers in a trunk meets industry standards; see Sec. 10.6.

10.5.2 Gain-frequency response

The gain of the amplifiers in typical amplifier/equalizer combinations is essentially constant at 20 dB from 50 to 450 MHz. The equalizer, however, is designed to reduce the gain of the combination to 6.9 dB at 50 MHz. This is because the loss of the cable for which the amplifier was designed increases from 6.9 dB at 50 MHz to 20 dB at 450 MHz, so that the frequency response of the transmission line/amplifier-equalizer combination is approximately constant over the entire frequency band.

10.5.3 Effect of cascading

The system noise amplitude is the root sum square of the noise amplitudes of the system components. The signal-to-noise and carrier-to-noise ratios are usually expressed in decibels, and the signal-to-noise ratio for n identical cascaded amplifiers, each with a signal-to-noise ratio of S_o/N_o, is

$$\frac{S_n}{N_n} = \frac{S_o}{N_o} - 10 \log n \tag{10.2}$$

A number of rules—some empirical, others based on theory—have been developed for estimating the cumulative noise and distortion of a number of amplifiers in series. If d_1, d_2,...are the distortions of successive elements in a system, the total distortion D may be the linear sum:

$$D = d_1 + d_2 + d_3 + \cdots \tag{10.3}$$

or follow a 3/2 power rule:

$$D = (d_1^{3/2} + d_2^{3/2} + \cdots)^{2/3} \tag{10.4}$$

or the root-sum-square rule:

$$D = (d_1^2 + d_2^2 + \cdots)^{1/2} \tag{10.5}$$

Differential gain and phase distortion increase in accordance with the 3/2 rule.

The frequency response of the system is the product of the responses of individual components.

10.5.4 Automatic gain and slope control

If a system has many amplifiers in series, a small change in the gain or frequency response of each amplifier resulting from a common cause such as a temperature change will produce a major change in the gain and frequency response of the system because of the compounding effect. To maintain the gain and frequency response of the system within close tolerances, inverse feedback is introduced by the addition of pilot tones near the bottom and top of the frequency band. Using these tones as reference, the gain and frequency response of each amplifier is automatically maintained at a stable level.

10.6 Trunk Circuits

Line amplifiers divide trunk circuits into segments, each having the length at which the attenuation at the highest transmission frequency is equal to the gain of the amplifiers. Equalizers are added that reduce the gain at lower frequencies so that the gain of each cable segment/amplifier/equalizer combination is 0 dB. As an example, assume the system bandwidth is 50 to 450 MHz, the gain of each amplifier is 20 dB, and the cable attenuation rises from 3.1 dB/1000 ft at 50 MHz to 10 dB/1000 ft at 450 MHz. The spacing between amplifiers would be (1000 ft)(20 dB/10 dB), or 2000 ft. An equalizer would be added to each amplifier with an attenuation varying from 6.9 dB at 50 MHz to 0 dB at 450 MHz. If the slopes of the equalizers are properly designed, the combined gain of the cable, amplifier, and equalizer would be 0 dB over the entire band.

10.7 Reverse Path Circuits

It is sometimes desired to bring signals from remote program sources back to the head end in a reverse direction. This can be accomplished on the frequencies below channel 2—6.0 to 30.0 MHz-where there is space for four 6-MHz channels. These channels have the advantage of lower attenuation but the disadvantage of higher ambient noise levels.

Special care must be exercised in the design of two-way systems. Every drop to a receiver is a noise source, and every return path must be blocked except the one carrying the desired signal. Otherwise the S/N of the system drops to a level that makes it unusable.

10.8 Video Performance Standards

As the cable TV industry has matured, various industry and governmental organizations have issued performance standards for CATV systems. Table 10.3 summarizes the basic performance standards that

TABLE 10.3 CATV Transmission Standards

	FCC 76.605	NCTA	EIA RS250B
Frequency response	± 2 dB	—	± 0.7 dB
Differential gain	—	10%	4%
Differential phase	—	5°	1.5°
Cross modulation	46 dB	—	—
Co-channel interference	36 dB	—	—
C/N (unweighted)	36 dB	—	—
S/N (weighted)	—	55 dB	56 dB

have been issued by the FCC, the NCTA, and the EIA for the transmission of NTSC signals. It illustrates the fact that there is not complete industry agreement on performance standards or on parameters that should be specified.

The standards for carrier-to-cross-modulation products and cochannel interference in part 76.605 of the FCC Rules are particularly important specifications for CATV systems because there are no guard bands between channels.

10.9   Head Ends

The head end of a CATV system collects the signals of all its program sources, processes them for transmission, generates and modulates channel carriers for signals that are in baseband form, shifts the channel frequency where necessary, and combines the resulting channels for transmission on the system.

10.10   Head-End Equipment

10.10.1   Broadcast channel antennas and receivers

The enormous expansion of CATV service has been made possible by satellite program distribution, but its backbone continues to be the distribution of standard broadcast signals. A CATV system head end, therefore, requires high-quality receiver-antenna combinations for off-air pickup.

The antenna must be located at a sufficient height to receive strong signals from the desired broadcast stations. It may be a commercial version of an all-channel home antenna, although the all-channel feature is often not satisfactory—one problem is that the off-air stations may be located in different directions—and most CATV systems use single-channel, high-gain antennas or antennas that receive only a few channels.

The most common antenna formats are the familiar *Yagi,* which is frequently used in home antennas, and the *log periodic,* in which each element is driven, thus giving better control over the antenna's directivity and bandwidth.

The principal components of the receiver are the preamplifier and signal processor. If the field strength is weak at the antenna location, the S/N of the system can be improved by placing a preamplifier at the antenna location. The signal at the output of the preamplifier is converted to the intermediate frequency for further amplification and signal processing. The IF signal is then heterodyned to the desired RF channel, where it is mixed with other signals for transmission and distribution over the system.

10.10.2 Video, microwave, and satellite signals

The signals from local program sources, microwave systems, and satellite systems are in the video and audio baseband formats, and after processing they are used to modulate channel carriers. The modulated signals should meet FCC standards for broadcast signals.

10.11 Auxiliary Radio Services

Three auxiliary radio services, Multipoint Distribution Service (MDS), Multichannel Multipoint Distribution Service (MMDS), and Cable Television Relay Service (CARS), are used to supplement and broaden the coverage of CATV systems.

10.11.1 Multipoint Distribution Service (MDS)

MDS is allocated 10 MHz of spectrum space, from 2150 to 2160 MHz. It is a point-to-multipoint service authorized to transmit single-channel video and data signals to customer-selected locations within a metropolitan area.

10.11.2 Multichannel Multipoint Distribution Service (MMDS)

MMDS is allocated 48 MHz (eight amplitude-modulated TV channels), from 2596 to 2644 MHz, to transmit TV programs and data to customer-selected locations. It is used for the distribution of television programs in sparsely populated areas where cable is uneconomic.

10.11.3 Cable Television Relay Service (CARS)

CARS is an inexpensive microwave service for transmitting a large number of CATV program signals from one site to another, as from head end to head end. The amplitude-modulated CATV channels from 112 to 300 MHz are heterodyned upward to 12,758.5 to 12,946.5 MHz and transmitted as microwave channels. This is called an amplitude-modulated link, or AML. For relatively short distances, an AML is usually more economic than cable or conventional microwave service.

10.12   HDTV Systems

Transmission standards have not been established (1991) for HDTV systems, and it is not possible to proceed with assurance in designing a CATV system for easy conversion to HDTV. The certainties are that HDTV cable systems should have greater bandwidth and a greater S/N, perhaps by 3 to 6 dB. These possibilities should be considered when designing the system for NTSC standards.

Satellite Systems

11.1 Satellites and Television

The use of satellites for CATV and television broadcasting program transmission and distribution began in the mid-1970s, and its subsequent growth has been phenomenal. Satellites are now indispensable to both industries for the transmission and distribution of video programs, and their utilization continues to grow.

The value of satellites in these functions results from four unique properties:

1. The cost of a satellite circuit is independent of its length. It costs no more to send a signal across the country than across the street. This property generally makes satellites cost-effective for long-haul, point-to-point service.

2. Except for a short segment at the beginning and end of each circuit, the path of the satellite signal is far above the surface of the earth and is unaffected by the nature or accessibility of the intervening terrain. This property makes satellites especially useful for paths over oceans, jungles, mountains, and arctic regions.

3. Once the satellite is in place, a new circuit can be established very quickly. This makes satellites valuable for *satellite news gathering* (SNG), where circuits must be set up quickly and with little preplanning. In this role they are of equal value for local and international news events.

4. Finally, satellites are especially adapted for point-to-multipoint service. With the satellite in place, for the relatively low cost of *receive-only earth stations* (see Sec. 11.6), service can be supplied to an indefinite number of earth stations. The most prominent use of this capability is the distribution of programs to CATV systems. It

is also used by the television broadcast networks for distribution of programs to their affiliates, and it probably will be used eventually for the distribution of programs directly to individual homes.

The importance of satellites to the television industry is indicated by the number of satellite channels that are devoted to the transmission of television programs in the United States. At the end of 1989 there were the following:

Service	Number of channels
CATV	122
Broadcast—full-time	20
Broadcast—part-time	146

The part-time channels in the broadcast service are leased on an *occasional use* basis, primarily for news and sporting events.

11.2 Satellite Communication Systems Elements

The elements of a satellite communication system are shown in Fig. 11.1. An earth station *uplink* transmits a signal to the satellite by means of a very narrow beam. Here it is received, amplified, and its frequency shifted to the *downlink* frequency by a *transponder*. The signal is retransmitted to earth by the *downlink,* covering an area known as the satellite's *footprint.*

The boundary of the footprint is the contour of *effective isotropic power,* EIRP, which is sufficiently high to provide satisfactory reception. The EIRP is the ratio in decibels between the actual signal strength and the signal from an isotropic antenna radiating one watt.

11.3 The Geosynchronous Orbit

11.3.1 Orbit location

The *geosynchronous orbit* is a circle in the *equatorial plane,* 22,300 miles above the earth's surface; see Fig. 11.2. It has the property that the centrifugal force acting upward on objects located on the orbit and rotating with the earth equals the gravitational force acting downward. Under these conditions, objects will appear to be stationary as viewed from the earth.

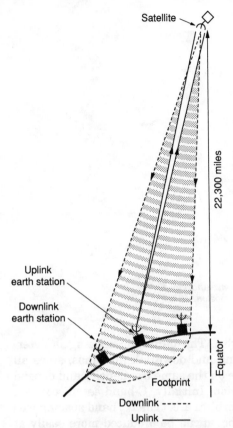

Figure 11.1 Satellite system elements.(*From Andrew F. Inglis*, Satellite Technology, *Focal Press, Boston, 1991.*)

11.3.2 Orbital slots

Satellites are assigned fixed positions on the geosynchronous orbit known as *orbital slots*, which are identified by their longitude. By international agreement, each country is allocated an arc of the geosynchronous orbit for its satellites, and the slots within this arc are assigned to individual licensees by the country's regulatory authority, the FCC for the United States.

The United States is allocated the arcs 62–103° and 120–146° west longitude for C-band satellites and 62–105° and 120–136° west longitude for the K_u band. (See sec. 11.4.3 for definitions of the C band and the K_u band.) The slots allocated to Canada are located in the gaps in the U.S. arcs.

The spacing between orbital slots is critical, and should be as small as possible to increase the number of satellites that can be accommo-

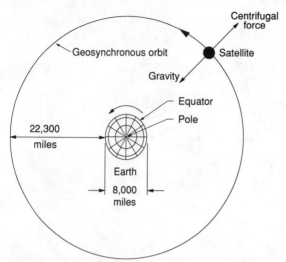

Figure 11.2 Geosynchronous orbit. (*From Andrew F. Inglis*, Satellite Technology, *Focal Press, Boston, 1991.*)

dated in the geosynchronous orbit. The minimum spacing is determined by the directivity of uplink and downlink antennas, since all satellites in a given band operate on the same frequencies and depend on antenna directivity for isolation. Initially, C-band satellites were spaced 4° and K_u band 3° on the orbital arc. (The K_u band spacing was less because a narrow antenna beam can be produced more easily at its shorter wavelength.) Later, both were reduced to 2°.

The spacing between DBS (direct broadcast by satellite) satellites will be wider, 9°, because it is assumed that smaller receiver antennas with wider beam angles will be used in this service.

11.3.3 Elevation and azimuth angles

The *elevation* or *look* angle of a satellite is the angle above the horizontal of the line of sight from a point on the earth to the satellite. The *azimuth* angle is the longitudinal position of the line of sight measured in degrees of west of Greenwich from 0° to 360°.

To calculate the elevation and azimuth angles from a point on the earth's surface, first calculate the path length L and the angle β:

$$\beta = \cos^{-1}(\cos \Delta \text{long} \cos \text{lat}) \tag{11.1}$$

$$L = (18.2 - 5.4 \cos \beta \times 10^{-4}\, \text{km} \tag{11.2}$$

where Δlong is the difference between the earth station and satellite longitudes and lat is the latitude of the earth station.

The elevation and azimuth angles are then

$$\text{Elevation angle} = \cos^{-1}(4.22 \times 10^4/L)(\sin \beta) \qquad (11.3)$$

$$\text{Azimuth} = (180° + \tan^{-1}(\tan \Delta\text{long/sin lat}) \qquad (11.4)$$

Assuming that there is a choice, it is desirable that a satellite be used that has as high an elevation angle as possible. With low elevation angles there may be difficulty in clearing obstacles, and the path through the troposphere is longer, which may result in more rain attenuation (especially in the K_u band), multipath distortions, and electrical noise generated by the earth's heat.

It is considered good design practice to utilize a minimum elevation of 5° for the C band and 10 to 20° for the K_u band.

11.3.4 Prime orbital arcs

The prime orbital arc for a location is the range of orbital slots that have an elevation angle in excess of 5°. The prime C-band arcs for CONUS (continental United States), Alaska, and Hawaii are:

	Degrees west longitude
CONUS	55–138
CONUS plus Hawaii	80–138
CONUS plus Anchorage	88–138

The prime arcs for K_u band are shorter because of the higher minimum angle. The arcs are also shorter at more northerly latitudes, and satellites are below the horizon at the north pole.

11.3.5 Solar eclipses

Solar eclipses occur when the earth comes between the satellite and the sun. They are not of direct interest to satellite users, but they have a major effect on satellite design. Satellites depend on *solar cells* for electric power, and they must be equipped with on-board batteries to supply power during the eclipses.

Eclipses occur near midnight at the satellite's longitude for a period of 21 days centered on the spring and autumn equinoxes. Their duration is a maximum of 70 minutes.

11.3.6 Sun outages

Sun outages occur when the sun and the satellite are in line as seen from a receiving location. The high temperature of the sun causes it to transmit a high-energy noise signal to earth station receiving antennas whenever it passes behind the satellite. The increase in noise is so great that a signal outage usually results.

The length and duration of the outages depends on the latitude of the earth station and the size of the receiving antenna. At an average latitude of 40° in the United States and with a 10-m antenna, outages occur for a period of 6 days with a maximum duration of 8 min.

Outage times for specific locations can be calculated or can be obtained from the satellite operator.

Sun outages cannot be avoided, and if continuous service is to be maintained, the traffic must be switched to another satellite.

11.4   Satellite Classifications

Satellites are classified by their *usage* and by their *technical* characteristics.

11.4.1   Usage classifications

The FCC defines two usage classifications of communication satellites. Fixed service satellites, FSS, is a broad classification that includes most commercial communication satellites. DBS, direct broadcast by satellite, is a specialized classification for satellites designed to broadcast directly to homes.

11.4.2   Technical classifications

The most important technical classifications of satellites are their spectrum bands and service areas.

11.4.3   Spectrum bands

Three spectrum bands are used for commercial satellite communications: C-band FSS, K_u-band FSS, and DBS. The uplink and downlink frequencies allocated to these bands are:

Service	Uplink	Downlink
C-band FSS	5.925–6.425 GHz	3.700–4.200 GHz
K_u-band FSS	14.0–14,5	11.7–12,2
DBS	17.3–17.8	12.2–12.7

The characteristics of satellite service in these three bands are quite different with respect to propagation effects, equipment performance, and FCC regulation.

The C and K_u bands have both advantages and disadvantages. C-band satellites share spectrum space with microwave systems. As a result the location and directional patterns of C-band earth stations must be *coordinated* or "cleared" with nearby microwave stations. It is sometimes impossible to clear a site in crowded metropolitan areas, and it is often not practical to use C-band uplinks for satellite news gathering where the uplink location cannot be planned in advance. In addition, the EIRP of C-band downlinks is limited to prevent interference to microwave systems.

C-band earth stations require larger antennas because of their shorter wavelength.

Offsetting these problems, the C band has important advantages, the most important being virtual immunity from signal outages resulting from rain attenuation. C-band components are generally lower in cost, in part because of larger mechanical tolerances.

11.4.4 Service areas

Satellite service areas vary widely in size, ranging from a metropolitan area to most of a hemisphere. The size of the service area depends on the size of the downlink antenna on the satellite—the larger the antenna, the narrower the beam and the smaller the service area. One classification defines four service area sizes, *global, regional, national,* and *spot.*

Global beams cover the entire area that is visible from the satellite. They are used for international communications.

Regional beams cover a group of countries, for example, western Europe.

National beams cover all or a significant portion of a single country. Most domestic U.S. satellites are in this category.

Spot beams cover a limited area and are used principally for point-to-point voice and data communications. They are seldom used in television service.

11.5 Satellite Construction

The construction of a communication satellite is shown in Fig. 11.3. It includes two primary systems, the *bus* and the *payload.*

The bus includes the satellite housing and the subsystems that provide power and maintain the satellite in its orbital slot.

The payload includes the communications subsystems—the antennas, the receivers, and the *transponders* that shift the frequency and amplify the uplink signals for retransmission.

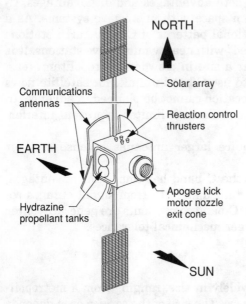

Figure 11.3 Communication satellite. (*From Andrew F. Inglis,* Satellite Technology, *Focal Press, Boston, 1991.*)

11.5.1 Satellite bus

The satellite housing can be rectangular, as shown in Fig. 11.3, or cylindrical, depending on the choice of the stabilizing mechanism that maintains the *attitude* of the satellite. The satellite in Fig. 11.3 employs internal gyroscopes and *three-axis stabilization.* An alternative is *spin stabilization,* in which the housing is a slowly rotating cylinder and the orientation of the antenna is held fixed through a rotating joint.

The power system consists of arrays of solar cells with battery backup for use during the solar eclipses. The solar cells can be mounted on flat panels as in Fig. 11.3 or on the cylindrical surface of a spin-stabilized satellite.

Although the forces on a geosynchronous satellite are nominally in balance, there are minor perturbations, e.g., the effect of irregularities on the earth's surface, that would cause it to leave its slot if uncompensated. Maintaining the satellite in its slot is known as *station keeping,* and it is achieved by ejecting small amounts of gas through the *reaction thrusters.* The amount of gas that can be stored in the satellite is one of the elements that limits its life.

11.5.2 Satellite payload

A functional diagram of the satellite's payload, its reason for being, is shown in Fig. 11.4. It includes the transmitting and receiving anten-

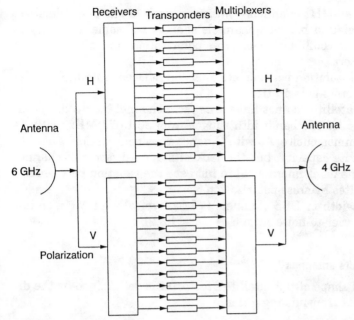

Figure 11.4 Satellite payload.

nas, the receivers, and the transponders. Separate channels are provided for vertically and horizontally polarized signals (see Sec. 11.5.3).

11.5.3 Frequency reuse

The principle of *frequency reuse* nearly doubles the satellite's communications capacity by employing both horizontal and vertical polarization (right- and left-handed circular polarization for some DBS systems). The *de facto* standard for the utilization of the downlink
spectrum of a C-band satellite is shown in Fig. 11.5. It includes

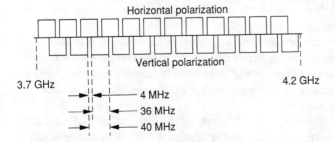

Figure 11.5 C-band satellite channels. (*From Andrew F. Inglis*, Satellite Technology, *Focal Press, Boston, 1991.*)

twenty-four 40-MHz channels—twelve in each of the two polarities. Adequate isolation between channels sharing the same frequency is achieved by a combination of cross-polarization and offsetting the channel centers.

Additional isolation between channels on adjacent satellites is also achieved by cross-polarization.

K_u-band satellites do not have a *de facto* channel bandwidth standard. In practice the bandwidth ranges from 36 to 72 MHz, with 54 MHz as a common choice. As with C-band, cross-polarization is used to double satellite capacity, but the lack of standardization of channel bandwidth makes it impractical to increase the isolation between adjacent satellites by cross-polarization.

The bandwidth of DBS channels is less, typically 24 MHz, to improve the carrier-to-noise ratio (see Sec. 11.8.2).

11.5.4 Satellite antennas

The size and shape of the satellite's footprint is determined by the directivity of its transmitting antenna.

11.5.5 Satellite receivers

The satellite normally receives a strong signal from the earth station, and the demands on the receiver's sensitivity are not great. There are normally two pairs of receivers in each satellite, two (one spare) for each group of channels with like polarization.

11.5.6 Satellite transponders

The transponder receives an IF signal for a single channel (sometimes two channels in K_u-band systems) from the receiver, shifts its frequency to the downlink channel, and amplifies it for retransmission to the earth. Transponders originally employed *traveling-wave tubes*, but, as with much of the electronics industry, a transition has been made to *solid-state power amplifiers,* SSPAs.

Most transponders exhibit the property of *saturation,* a power level above which an increase in input causes no further increase in output. The output of the receiver should drive the transponder to saturation.

Transponder power is determined by FCC regulations, economics, and technical feasibility.

In the C band, the FCC (and international) limit on downlink power density limits the transponder power to about 8 W. The power density limit is stated in terms of power per kilohertz of bandwidth per unit area. For FM signals with energy dispersal (see Sec. 11.6.1), it translates to an EIRP of from 38 dBw at elevations below 25° to 48 dBw at higher elevation angles.

The energy density limit is much higher in the K_u band, and economics and technical feasibility establish the limits on transponder power; 40 W is a typical choice.

Even higher power is required for DBS, and transponder powers up to 250 W and more have been proposed.

11.6 Earth Stations

Earth stations come in a wide range of size, cost, and performance. *Homesats,* small receive-only stations (TVROs) that sell for less than $2000 and are designed for home reception, are at one end of the range. At the other end are multimillion-dollar installations owned by satellite operators and program suppliers.

11.6.1 Earth station antennas

The antenna is a basic component of both uplink and downlink earth stations. Most earth station antennas employ the familiar format of backyard TVROs, a parabolic or quasi-parabolic reflector *illuminated* by a feed horn. Their performance, however, differs widely depending on their size and design.

The directional properties of earth station antennas are shown in Fig. 11.6. They are rated by their gain with respect to an isotropic radiator, their half-power beamwidth (HPBW), their null beamwidth (NBW), and their side-lobe amplitude.

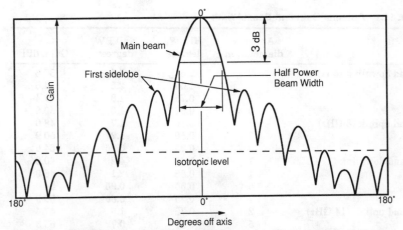

Figure 11.6 Earth station antenna radiation pattern. (*From Andrew F. Inglis, Satellite Technology, Focal Press, Boston, 1991.*)

The shape of the pattern depends on the size of the antenna and the uniformity of the illumination of the antenna aperture by the feed horn. The width of the beam is inversely proportional to the antenna and is a minimum when the illumination is uniform. Uniform illumination results in rather large side lobes, however, and these can be reduced at the expense of greater beamwidth by tapering the illumination from the center to the edge of the reflector.

The half-power beamwidth is given by Eq. (11.5):

$$\text{Half-power beamwidth} = C_{\text{HP}}\left(\frac{\lambda}{D}\right) \qquad (11.5)$$

where C_{HP} varies from 50° for uniform illumination to 70° for typical practical designs and D is the antenna diameter.

The null beamwidth is given by Eq. (11.6):

$$\text{Null beamwidth} = C_{\text{null}}\left(\frac{\lambda}{D}\right) \qquad (11.6)$$

where C_{null} varies from 114° to 170°.

The gain of the antenna is inversely proportional to the square of the beamwidth (or the product of the vertical and horizontal beamwidths if different). It is also affected by the antenna efficiency, the fraction of the energy that is radiated after subtracting ohmic losses and energy intercepted by the feed horn. Table 11.1 lists specifications of representative uplink and downlink antennas, assuming 70 percent efficiency and nonuniform illumination.

TABLE 11.1 Antenna Gain and Beam Width

	Antenna diameter, m	HPBW, degrees	Null BW, degrees	Gain, dBi
C-band downlink (4 GHz)	2	2.6	6.4	36.6
	5	1.05	2.5	44.5
	7	0.75	1.8	47.4
	10	0.52	1.3	50.6
C-band uplink (6 GHz)	5	0.70	1.7	48.0
	7	0.50	1.2	50.9
	10	0.35	0.85	54.0
K$_u$-band downlink (12 GHz)	1	1.75	4.2	40.0
	1	0.88	2.1	46.0
	5	0.35	0.95	54.0
	7	0.25	0.60	56.9
K$_u$-band uplink (14 GHz)	2	0.75	1.8	47.4
	5	0.30	0.72	55.3
	7	0.21	0.51	58.4

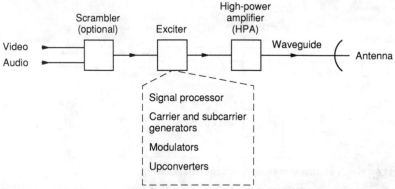

Figure 11.7 Earth station uplink functional diagram. (*From Andrew F. Inglis*, Satellite Technology, *Focal Press, Boston, 1991.*)

11.6.2 Uplink functional diagram

A functional diagram of a typical earth station uplink is shown in Fig. 11.7.

The video and audio signals may be *scrambled* to prevent unauthorized reception.[1] They then pass to the exciter, the heart of the earth station. It generates the intermediate-frequency carrier, typically 70 MHz for the visual carrier, and adds *preemphasis* (see Sec. 11.8.4) and a *sideband energy dispersal* waveform (see Sec. 11.8.6) that modulates the IF carrier and *up-converts* it to the final carrier frequency, which drives the power amplifier.

The power of the HPA in combination with the uplink antenna gain should be sufficient to drive the satellite transponder to saturation. This depends on the characteristics of the satellite, but a typical figure is 83 dBw. If the antenna gain were 54 dB (see Table 11.1), the required power delivered to the antenna would be 29 dBw and a rated power of 1 kW would probably be chosen.

Traveling-wave tubes are frequently used for HPAs, but there is a trend to solid-state devices.

Uplink antennas must meet the FCC specification shown in Table 11.2, which limits the off-axis gain of uplink antennas to avoid interference to adjacent satellites.

11.6.3 Downlink functional diagram

A functional diagram of a satellite downlink earth station is shown in Fig. 11.8.

11.6.4 Downlink signal-to-noise ratio

The most significant performance specification of a satellite link is its signal-to-noise ratio S/N, and this is determined to a large extent by

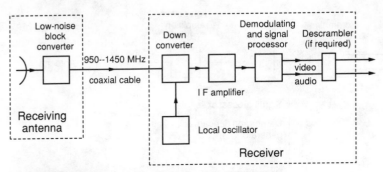

Figure 11.8   Earth station downlink functional diagram.

TABLE 11.2   Maximum Off-Axis Radiation

Off-beam angle, degrees	Maximum gain, dBi
1	29.0
2	21.5
3	17.0
4	14.0
5	11.6
6	9.6
7	8.0

the performance of the downlink earth station. See Sec. 11.9.2 for the S/N calculation.

A satisfactory signal-to-noise ratio is more difficult to achieve with satellites than with microwave circuits because of their long path lengths. On the other hand, microwave circuits are more subject to atmospheric fading, and the fade margin is a more important specification for them.

11.6.5   *G/T*, the downlink figure of merit

Other factors being equal, the signal-to-noise performance of a downlink earth station is proportional to the ratio G/T, where G is the gain of the antenna (see Table 11.1) and T is the *equivalent noise temperature* of the system in Kelvins, converted to decibels relative to 1 K.

Most of the noise in a satellite system is *random* and has the same characteristics as *thermal* noise, the electrical signals generated by a hot object. The equivalent noise temperature is a convenient measure of the noise contribution of a component or system, even though it is nonthermal in origin.

T is given by the equation

$$T = 290°(F_n - 1) \tag{11.7}$$

F_n is the *noise figure* expressed as an arithmetic (not logarithmic) ratio and is equal to the quotient of the signal-to-noise ratio at the output of a system divided by the S/N at the input normalized to an input temperature of 293°.

Because of its universality, the noise temperature can be used to specify the noise contribution of all the components of a system. The noise temperature of a system is then equal to the sum of the noise temperatures of its components.

The major sources of noise in practical systems are the antenna and the input stage of the receiver. The antenna noise is the noise power radiated from the earth because of its warmth and picked up by the antenna. The input stage of the receiver is a *low-noise amplifier*, LNA, chosen for its low equivalent noise temperature.

The amount of electrical noise power received from the earth by the antenna and its equivalent noise temperature depend on its directivity and its elevation angle. Noise also originates in heavy rainfall, which is also a source of noise in the K_u band.

Typical antenna noise temperatures are shown in Table 11.3.

TABLE 11.3   Antenna Noise Temperature (K)

Elevation angle	C band	K_u band
5°	58	80
10°	50	58
30°	30	38

Important progress has been made in recent years in the development of reasonably priced LNAs. Among these are the field-effect transistor (FET) and the parametric amplifier. For the most demanding applications, the parametric amplifier can be artificially cooled. Equivalent noise temperatures of typical types of these LNAs are shown in Table 11.4.

The LNA is mounted at the antenna and provides the initial amplification. Alternatively, as in Fig. 11.8, a *low-noise block converter* amplifies and translates all the incoming signals to a lower frequency range, often 950 to 1450 MHz, for transmission to the receiver by coaxial cable.

The remainder of the receiver functions are conventional.

TABLE 11.4   LNA Noise Temperatures (K)

LNA type	C band	K band
Field-effect transistor	85	150
Parametric amplifier	60	120
Cooled parametric amplifier	29	43

11.7 Transmission Formats

The transmission mode used by satellite systems depends on the format of the baseband signal.

Frequency modulation is more commonly used for the transmission of analog video signals, whether in the component or composite format. Its advantages are relative noise immunity, insensitivity to system nonlinearities, and the ability to spread the transmitted energy over the entire channel by *sideband energy dispersal*; see Sec. 11.8.6. Its disadvantage as compared with amplitude modulation is its greater bandwidth requirement. Its performance as compared with digital transmission is described below.

11.7.1 Frequency modulation

The instantaneous voltage waveform and the sideband spectrum for a carrier frequency modulated with a video signal are shown in Fig. 11.9. The sync tip is clamped to a reference frequency so that the dc component of the signal is retained.

11.7.2 Digital modulation

Many modulation modes are available for the transmission of digital signals. They employ *amplitude-shift keying, phase-shift keying,* or *frequency-shift keying,* either singly or in combination. Each mode of-

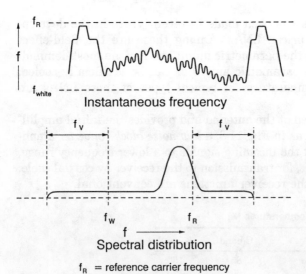

f_R = reference carrier frequency
f_V = maximum video frequency

Figure 11.9 Frequency-modulated video signal.

fers a different tradeoff between required power per bit, maximum signal speed, sensitivity to fading, and sensitivity to co-channel interference. See Sec. 4.4 for a comparison.

11.8 FM Performance Characteristics

11.8.1 Bandwidth requirements

The bandwidth of a frequency-modulated signal is greater than the peak-to-peak frequency deviation and is determined by the spectrum of its sidebands. In theory they extend to infinity in both directions, but almost all of the sideband energy is confined to a frequency band extending from the extremes of the instantaneous frequency deviation waveform by an amount approximately equal to the video baseband bandwidth. The approximate total IF (or RF) bandwidth B is

$$B_{IF} = \Delta f_V + 2B_V \qquad (11.8)$$

where Δf_V = peak-to-peak frequency deviation
$\quad B_V$ = video baseband bandwidth

The same frequency units should be used for all terms.

A complete 40-MHz C-band transponder is usually used for each video signal, although two could be compressed into this spectrum space at the expense of S/N. It is common to use a single 54-MHz K_u-band transponder for two video channels.

11.8.2 FM improvement factor

The *FM improvement factor* F_{FM} is a major advantage of the FM transmission mode. It is the ratio (in decibels) of the S/N of the demodulated receiver output to the *carrier-to-noise* ratio of the modulated carrier, usually measured at the receiver IF. It can be calculated by Eq. (11.9):

$$F_{FM} = 20 \log \left(\frac{\Delta f_V}{B_V} \right) + 10 \log \left(\frac{B_{IF}}{B_V} \right) + 7 \qquad (11.9)$$

where F_{FM} = FM improvement factor, dB
$\quad B_{IF}$ = bandwidth of the radio-frequency channel
$\quad \Delta f_V$ = peak-to-peak frequency deviation
$\quad B_V$ = video bandwidth

The ratio $\Delta f_V/f_V$ is the *modulation index*.
F_{FM} can be quite large, sometimes more than 20 dB.

11.8.3 Noise spectrum

An additional advantage of frequency modulation for video signals is the fact that the noise spectrum is triangular, i.e., the amplitude of the noise components increases linearly with frequency. Since the acuity of the eye is less for the fine-grained high-frequency components, the noise weighting factor is about 4 dB greater for FM noise deemphasis than for flat noise; see Sec. 2.16.3.

11.8.4 Frequency preemphasis/deemphasis

The signal-to-noise ratio of an FM transmission system can be improved further by the use of frequency preemphasis/deemphasis. The amplitude of the higher-frequency components of the video baseband is increased or emphasized for transmission and deemphasized at the receiver. The deemphasis reduces the amplitude of the high-frequency noise components.

Standardization is required to obtain the full benefits of preemphasis/deemphasis and the CCIR has issued Recommendation 405 (Table 11.5) for this purpose. It results in an 11-dB improvement in the signal-to-noise ratio of an NTSC signal transmitted by a frequency-modulated system.

11.8.5 Noise weighting factor

The noise weighting factor for video signals, which converts unweighted to weighted noise, is described in Sec. 2.16.3. The weighted signal-to-noise ratio for an NTSC signal transmitted by a frequency-modulated carrier, calculated by the EIA noise standard weighting curve, is about 12 dB higher than the unweighted value.

11.8.6 Sideband energy dispersal

Sideband energy dispersal is achieved by adding a low-frequency, 30-Hz (25-Hz for PAL), sawtooth waveform to the video signal to cycli-

TABLE 11.5 CCIR Preemphasis/Deemphasis
Standard, Recommendation 405-1

Baseband frequency	Relative response, dB
10kHz	−10
20	−10
50	−9.5
100	−8.8
200	−6.8
500	−2.0
1MHz	+1.4
2	+2.8
5	+3.6

cally move the range of the frequency deviation of the carrier. This spreads the sideband energy over a larger portion of the spectrum and prevents the heavy concentration in one area that would severely limit the amount of downlink power because of the FCC limitation on energy density.

11.9   Satellite Link Calculations

11.9.1   Design parameters

The design of a satellite link for analog signals requires the calculation of its signal-to-noise ratio and its *fade margin.*

The design of a satellite link for digital signals requires the calculation of its bit error rate. Usually the link contributes only a small amount to total system noise until the ber exceeds the critical value where the error correction becomes ineffective. Noise above this level makes the signal virtually unusable.

With typical error correction circuits, the ber will be below this level on links that have a satisfactory signal-to-noise ratio for analog signals. The reader should consult the references for a more precise calculation.

11.9.2   Signal-to-noise ratio

The signal-to-noise ratio of a satellite link using frequency modulation is calculated by determining its *carrier-to-noise* ratio and converting this figure to the signal-to-noise ratio by means of Eq. (11.10).

$$\frac{S}{N} = \frac{C}{N} + F_{FM} + F_{DEEMPH} + F_W \qquad dB \qquad (11.10)$$

where F_{FM} is the *FM improvement factor.* It is the ratio of the S/N of the demodulated video baseband to the C/N of the unmodulated carrier. F_{FM} (dB) is

$$F_{FM} = 20 \log \left(\frac{\Delta f_V}{B_V}\right) + 10 \log \left(\frac{B_{IF}}{B_V}\right) + 7 \text{ dB} \qquad (11.9)$$

where B_{IF} is the IF bandwidth expressed in the same units as $\Delta(f_V)$ and B_V; and the term $\Delta f_V/B_V$ is the *modulation index m.*

C/N is the carrier-to-noise ratio, the ratio of the power of the unmodulated carrier to the noise power. It is given by the equation

$$\frac{C}{N} = (\text{EIRP} - 20 \log f_{GHz} - 10 \log B_{Hz} + 45) + \left(\frac{G}{T}\right)_{dB/K} \qquad (11.11)$$

where EIRP is the effective isotropic power of the satellite footprint at the receiver location (the EIRP at a given location can be obtained from the satellite operator), f_{GHz} is the carrier frequency in *GHz*, and B_{Hz} = IF bandwidth in *Hz*. For $(G/T)_{dB/K}$, see Sec. 11.6.5.

Note that C/N is inversely proportional to the bandwidth, an important factor in system design.

11.9.3 Fade margin

The fade margin is the difference (in decibels) between the receiver input power under nonfading conditions and at its *threshold*. The threshold is a property of FM receivers and is the point at which the signal-to-noise ratio, S/N, of the output signal begins to drop much more rapidly than the C/N. The fade margin is given by the equation

$$\text{Fade margin} = P_R - T_S \tag{11.12}$$

where P_R = power at the terminals of the receiver under nonfading
        conditions
    T_S = threshold sensitivity of the receiver

The input power to the receiver at that point is its *threshold sensitivity*.

Since the signal quality decays rapidly at power levels below the threshold, it is important that the fade margin be sufficient to maintain service, even during fading from refractive effects or heavy rainfall.

The length of the path of satellite signals through the troposphere is relatively short, and tropospheric fading is not a severe problem except for K_u-band signals in regions of torrential rainfall. For C-band and many K_u-band satellite circuits, a 3-dB fade margin is adequate.

Many receivers have *threshold extenders* that reduce the threshold power, typically to the range of 11 to 8 dBm.

11.10 S/N and Fade Margin of Representative Satellite Systems

This section illustrates the principles that establish the S/N and fade margin of satellite links by applying them to representative examples.

1. A C-band trunk circuit of highest quality and reliability such as would be used for interconnecting two major switching centers.

2. A C-band downlink to a CATV head end.

3. A K_u-band ENG circuit; one-half of a 54-MHz transponder is assumed.

4. A C-band homesat.

5. A K_u-band homesat operating with a higher EIRP than is permitted on the C band. This is sometimes called medium-power DBS.

6. High-power DBS.

7. An HDTV circuit.

The critical circuit parameters that have been assumed for these examples are shown in Table 11.6, together with the resulting signal-to-noise ratios, weighted and unweighted, and the fade margins. A receiver threshold of 8 dB was assumed, except for the homesat and DBS downlinks, where 3 dB of threshold extension was assumed.

11.10.1   Trunk circuits

Trunk circuits should be designed to add negligible noise to the signal, since they are in series with other system elements, e.g., CATV downlinks, which for economic reasons leave very little margin for noise in the trunks. This is recognized in EIA/TIA-250-C performance standards, which specify a minimum *unweighted* S/N of 56 dB for point-to-point satellite transmission circuits. The example in Table 11.6 meets this standard.

Since the EIRP is limited by FCC Rules, a high S/N must be achieved with a receiving system having a high G/T—in this example, with a 10-m antenna and a cooled parametric amplifier for the LNA. This not only increases the S/N but permits a full bandwidth IF amplifier to be employed without adversely affecting the C/N and the fade margin.

The fade margin is 19 dB even without a threshold extender.

TABLE 11.6   Representative Circuit Parameters

	Trunk	CATV	ENG	Homesat	Homesat	DBS
Band	C	C	K_u	C	K_u	DBS
Downlink EIRP	38	38	43	38	43	50
Antenna						
Diameter, m	11	4.5	5	2	1	0.5
Gain, dB	55	42	54	36	40	34
Noise temperature, K						
Antenna	40	40	50	40	50	50
LNA	50	85	150	125	150	150
G/T, dBK	35	21	31	14	17	11
B_{IF}, MHz	32	32	24	24	24	27
C/N, dB	31	17	24	11	10	12
Bv, MHz	4.2	4.2	4.2	4.2	4.2	4.2
Δf_v, MHz	10.5	10.5	7.5	10.5	10.5	10.0
S/N, dB	62	48	52	42	38	44
Receiver threshold, dBm	12	8	12	8	8	8
Fade margin, dB	19	9	12	6	8	6

11.10.2   CATV downlinks

The downlinks are a significant part of the cost of a CATV system, and their G/T is usually chosen so that their S/N exceeds industry standards for system performance; see Sec. 10.5.1. With an unweighted S/N of 50 dB, the performance of the downlink in the example in Table 11.6 exceeds this standard. With the prospect of more stringent requirements for HDTV in the future, it is likely that the S/N standard for CATV downlinks will be increased.

The fade margin of 11.4 dB is adequate, even without the use of a threshold extender.

11.10.3   ENG circuits

A comparison of the CATV and ENG system components in Table 11.6 illustrates some of the differences between C-band and K_u-band performance parameters.

The gain of the K_u-band antenna is 8 dB higher because of the shorter wavelength, but this is offset by a greater space loss and a higher antenna noise temperature. The higher C/N of the K_u-band system results from the greater EIRP of its footprint and its lower IF bandwidth, the latter chosen to increase the fade margin to compensate for rain attenuation.

The fade margin of 15.6 dB is adequate except in areas of extremely heavy rainfall.

11.10.4   C-band homesats

The table clearly shows the compromises that must be made to operate a C-band homesat with a small antenna.

The IF bandwidth is reduced to increase the fade margin, but at the expense of frequency response. Even with this increase, the use of a threshold extender is desirable.

11.10.5 K_u-band homesat

The use of the FSS K_u-band, sometimes known as medium-power DBS, for direct-to-home transmission has been proposed. The higher power EIRP that is permitted, would permit the use of smaller downlink antennas. The example shown in Table 11.6 assumes not only a smaller 1-m antenna but also two video channels in a single transponder. The 42.2-dB weighted S/N is marginal, as is the 4.6-dB fade margin. This example, however, suggests the limits of medium-power DBS.

11.11   DBS Transmission Requirements

The DBS systems that have been proposed involve different combinations of transponder EIRP and downlink receiver G/T. Table 11.6

shows one example in which the EIRP is 56 dB (as compared with 45 dB for a typical K_u-band FSS satellite). With the assumptions used in the example, the fade margin would be 4.7 dB and the weighted S/N would be 45 dB. This is not excellent picture quality, but it is satisfactory for most uses.

11.12   HDTV Transmission Requirements

The requirements that will be placed on satellites for the transmission and distribution of HDTV signals are not known with precision or in detail at the present time because HDTV standards for spectrum utilization have not been adopted. A description of the possible usages for satellites in HDTV is contained in Chap. 14.

References

1. A. F. Inglis, *Electronic Communications Handbook*, McGraw-Hill, New York, 1988, Chap. 17.

Fiber Optic Transmission Systems

12.1 Overview

Fiber optics is one of the several new technologies that are revolutionizing the television industry. Because its transmission medium is infrared radiation, which in electromagnetic terms has an extremely high frequency, it is possible to construct fiber optic transmission systems with enormous bandwidths—20 GHz is typical. This bandwidth makes them especially well adapted for digital signals, although analog transmission is possible as well (see Sec. 12.5.2).

The cost of fiber optics now limits its applications to point-to-point transmission and CATV trunks (see Chap. 10), but as its technology advances and its cost is reduced, it can be expected that it will eventually be used for distributing voice, message, video, and audio service to individual homes.

Fiber optic systems are simple in principle, as shown in Fig. 12.1. An optical energy source, a laser or LED, is modulated with the video signal. It illuminates the end of an optical fiber, and the energy is transmitted by internal reflections within the fiber to its opposite end.

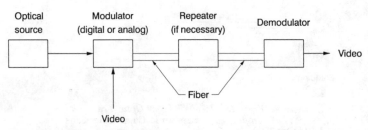

Figure 12.1 Fiber optic system.

Here it is detected and demodulated, and the original baseband signal is recovered.

Unlike in electromagnetic transmission systems, where it is unnecessary to demodulate and remodulate the signal at repeater points, demodulation and remodulation is required for fiber optic systems. This is not a serious problem for digital transmission, but analog transmission is used only on single-hop systems.

12.2   Fiber Optic Construction and Operation

12.2.1   Construction

A fiber (see Fig. 12.2) consists of an exceedingly fine glass filament that is surrounded by a concentric *cladding,* also made of glass but having a slightly lower index of refraction. The index-of-refraction profile for single-mode fiber is known as the *step index.* The index for the multimode fiber in Fig. 12.1 is carefully shaped because it determines the bandwidth and frequency response of the fiber; see Sec. 12.4.2. It is known as *graded-index* multimode fiber.

The diameter of the fibers depends on the transmission mode, see Sec. 12.3. The cores of *single-mode* fibers typically have diameters of 7 to 10 μm, while *multimode* cores range from 50 to 85 μm. One standard for multimode fibers is a 50-μm core and a 125-μm cladding. These diameters are comparable with those of human hair.

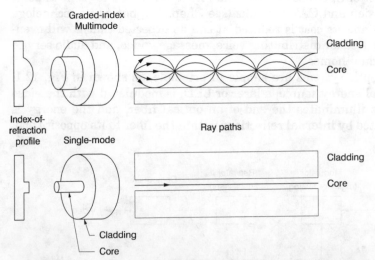

Figure 12.2   Fiber construction. (*From Ira Jacobs, "Fiber Optic Transmission Systems," chap. 8 in A. F. Inglis (ed.),* Electronic Communications Handbook, *McGraw-Hill, New York, Fig. 8.1.*)

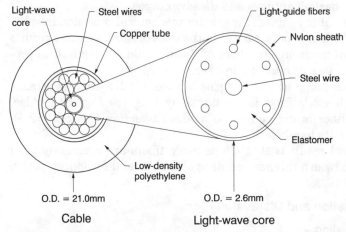

Figure 12.3 Fiber optic cable. (*From Donald G. Fink (ed.)*, Electronic Engineers' Handbook, *McGraw-Hill, New York, 1989, Fig. 22.25a.*)

A fiber optic cable consists of one or more fibers bundled together and surrounded by a metallic or plastic sheath for mechanical protection. The bundle may include a steel wire for mechanical strength; see Fig. 12.3. In one configuration, 12 groups, each containing 12 fibers, are bundled together. Needless to say, the bandwidth and hence the communication capacity of such a cable is enormous.

12.2.2 Operation

Infrared radiation is injected into the end of the fiber and is totally reflected when it strikes the cladding, with its lower index of refraction. It is contained within the core and continues down the fiber until it is completely attenuated by losses in the glass.

12.3 Single- and Multimode Fibers

The operation of single-mode and multimode fibers is illustrated in Fig. 12.2. The single-mode fiber has a smaller diameter, and only one mode, a single path down the fiber, is possible. The *shortest* wavelength that can be transmitted in a single mode is

$$\lambda_{\min} = 2.6r(n_1^2 - n_2^2) \tag{12.1}$$

where r is the radius of the core and n_1 and n_2 are the indices of refraction of the cladding and core.

Larger-diameter fibers permit multiple reflections, and multimode operation is possible.

Each mode has advantages and disadvantages.

Single-mode fiber requires far greater mechanical precision in splicing. A lateral offset of only 1.5 μm in the ends of the fiber cores results in a 0.5-dB attenuation. An offset of 7 μm would be required to produce the same loss in multimode fiber.

On the other hand, at wavelengths in excess of 1.1 μm, the attenuation of single-mode fiber is substantially less; see Fig. 12.4. Also, single-mode fiber is not subject to *modal dispersion* (see Sec. 12.4), which reduces its bandwidth.

The practical result is that single-mode fibers are commonly used for wider-bandwidth (higher-speed) applications and for longer paths.

12.4 Attenuation and Dispersion

12.4.1 Attenuation

Both single-mode and multimode fibers suffer from attenuation, as shown in Fig. 12.4 for a representative glass type. For this glass, the attenuation of single-mode fibers reaches a low of about 0.2 dB/km at wavelengths in the 1.5–1.6-μm infrared range. Long distance signal transmission by fiber optic cables was made possible by the development of energy sources in this region of the spectrum, together with the availability of glass with lower attenuation and the development of techniques for splicing the tiny single-mode fibers.

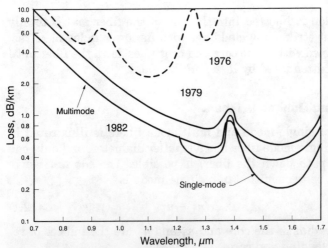

Figure 12.4 Fiber optic attenuation. (*From Ira Jacobs, "Fiber Optic Transmission Systems," chap. 8 in A. F. Inglis (ed.),* Electronic Communications Handbook, *McGraw-Hill, New York, Fig. 8.2.*)

12.4.2 Modal dispersion

The dispersion encountered in fiber optic circuits results from differences in the arrival time of different rays or different frequency components. It can be caused by differences in the path length of different modes, *modal dispersion,* or by differences in transmission speed for different wavelengths, *chromatic dispersion.*

The effect of dispersion is to cause pulses to spread, and this limits the maximum pulse rate. The effect can also be described as a bandwidth limitation. The relationship between bandwidth and pulse spreading is given by Eq. (12.2):

$$B = \frac{0.44}{\tau} \qquad (12.2)$$

where B = bandwidth, MHz

τ = width, μs, between the points on the edges of the pulse where it is one-half its maximum value

The bandwidth of multimodal fiber, when determined by modal dispersion, is a function of the wavelength and the index-of-refraction profile. The profile can be adjusted so that the index of refraction is greatest and the transmission velocity is the least on the shortest paths. This partially equalizes the transmission time, reduces the pulse spread, and increases the bandwidth. Figure 12.5 shows three

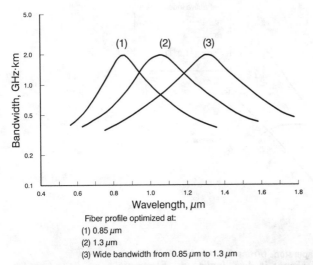

Fiber profile optimized at:
(1) 0.85 μm
(2) 1.3 μm
(3) Wide bandwidth from 0.85 μm to 1.3 μm

Figure 12.5 Fiber optic bandwidth. (*From Ira Jacobs, "Fiber Optic Transmission Systems," chap. 8 in A. F. Inglis (ed.),* Electronic Communications Handbook, *McGraw-Hill, New York, Fig. 8.3.*)

examples of bandwidth vs. wavelength for different index-of-refraction profiles.

12.4.3 Chromatic dispersion

Chromatic dispersion affects both single-mode and multimode fibers, and it is the principal bandwidth limitation for single-mode fibers. It occurs because fiber optic light sources are not monochromatic, and the transmission velocity varies with the wavelength.

Two effects are present that affect the variation of transmission velocity with wavelength. The first is *material dispersion* due to the variation in the refractive index of the fiber with wavelength. The second is *waveguide dispersion* that is caused by a graded index fiber—one whose index of refraction varies across its diameter. Material and waveguide dispersion are in opposite directions, and the waveguide dispersion can be controlled so that the variation in delay is equal to zero at some wavelength. The wavelength at which this occurs can be adjusted within limits by appropriate choice of the refraction index profile.

Figure 12.6 compares the dispersion in picoseconds per kilometer for each nanometer of source spectral width for two fibers, one with a uniform index of refraction across the core and the other with an index

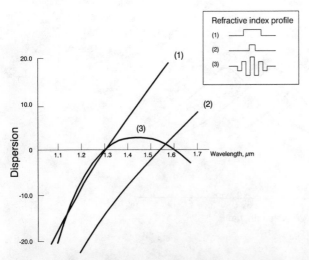

Figure 12.6 Spectral dispersion. (*From Ira Jacobs, "Fiber Optic Transmission Systems," chap. 8 in A. F. Inglis (ed.),* Electronic Communications Handbook, *McGraw-Hill, New York, Fig. 8.4.*)

profile that has been adjusted to move the zero dispersion wavelength from 1.3 μm to 1.6 μm.

12.5 Signal Formats and Modulation Modes

12.5.1 Digital format

The digital format is uniquely suited to the characteristics of fiber optics. Fiber optic systems have a surplus of bandwidth but are limited with respect to power and receiver sensitivity. *Pulse code modulation,* PCM, trades bandwidth for signal-to-noise ratio, and is almost universally used except for short single-hop paths.

Digital video signals with any of the standard bit rates described in Chap. 4 can be transmitted by fiber optics. They generally employ a simple binary code (see Chap. 4) with on-off keying, OOK. Other modulation modes such as quaternary pulse position modulation, QPPM, can be used to effect a different tradeoff between S/N and bandwidth.

Modes involving phase-shift or frequency-shift modulation are not practical for use with an optical carrier.

12.5.2 Analog format

The analog format is more economical because A-to-D and D-to-A conversion are not required, and it is practical for some single-hop systems. Amplitude modulation can be used with intensity modulation of the optical source. Alternatively, a subcarrier that is frequency-modulated by the signal can be caused to intensity modulate the light source. This eliminates problems of linearity and provides an increase in signal-to-noise ratio by the FM improvement factor; see Sec. 11.9.2.

Frequency modulation has the additional advantage that several video signals can be multiplexed on the same fiber by the use of FDM-FM, frequency division multiplex.

The range of a single-hop system is sometimes known as its *reach,* which is established by the signal-to-noise ratio. This calculation for multichannel FDM-FM systems is complex and beyond the scope of this book. The reach of these systems is suggested by Fig. 12.7,[1] which shows the S/N based on empirical results.

12.6 Path Performance, Digital Format

The measure of performance for a digital transmission system is the bit error rate rather than the signal-to-noise ratio as in analog systems. This is determined on a statistical basis and is expressed as a *bit error probability*.

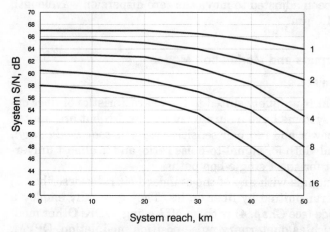

Figure 12.7   Range of analog systems. (*From Ira Jacobs, "Fiber Optic Transmission Systems," chap. 8 in A. F. Inglis (ed.),* Electronic Communications Handbook, *McGraw-Hill, New York, Fig. 17.19.*)

Bit errors can be caused not only by system noise, but also by *signal-dependent noise,* including such noise sources as intersymbol interference and improper setting of decision thresholds. In total, these are known as *eye degradation.* The calculation of bit error probability is beyond the scope of this book, and the reader should consult the references.

The ultimate criterion of the performance of fiber digital transmission systems is the required repeater spacing for multihop systems. This is not an exact calculation because it must be based on statistical results and it will depend on the performance standards. Approximate maximum spacings for single-mode and multimode fibers with laser and LED optical sources (see Sec. 12.7) are shown in Fig. 12.8. This graph clearly shows the increase in range that has resulted from the development of laser sources in the 1.3- to 1.6-μm range and the introduction of single-mode fibers.

12.7   Transmitters for Fiber Optic Systems

Two types of optical transmitters are used in fiber optic systems, lasers and light-emitting diodes (LEDs). Both are semiconductors, but there is a fundamental difference in that the radiation from lasers is *coherent,* that is, it retains its wavelength and phase over a period of time.

The use of lasers was limited at first because they were not available in the 1.5- to 1.67-μm wavelength range, the area of the lowest attenuation (see Fig. 12.4). The development of lasers that operated at these wavelengths was a significant breakthrough in fiber optic technology.

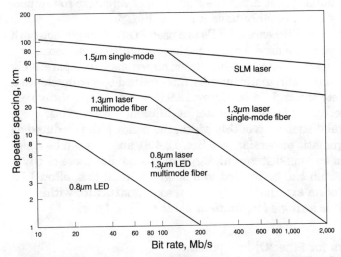

Figure 12.8 Repeater station spacing. (*From Ira Jacobs, "Fiber Optic Transmission Systems," chap. 8 in A. F. Inglis (ed.),* Electronic Communications Handbook, *McGraw-Hill, New York, Fig. 8.17.*)

Some important comparisons of lasers and LEDs are summarized in Table 12.1.

In these examples, both emit the same output power, but the coupling of the power into the fiber is far more efficient for lasers.

The spectral width of laser emission is much narrower, an important advantage because it minimizes the loss of bandwidth due to chromatic dispersion.

The maximum rate at which lasers can be modulated is greater by a ratio of about 10:1. The maximum rate for LEDs, approximately 500 Mbits/s, however, is high enough for most purposes.

The output power of lasers can be stabilized by feedback, an operation that is not possible with LEDs.

TABLE 12.1 Lasers and LEDs

Parameter	Laser	LED
Light output	6 dBm	6 dBm
Coupling loss	3 dB	20 dB
Spectral width		
At 0.8 μm	2 nm	40 nm
At 1.3 μm	4 nm	100 nm
Maximum modulation rate (approx.)	5 Gbs	500 Mbs
Feedback power control	Yes	No
Temperature sensitivity	Strong	Weak
Failure mechanisms	Many	Wears out

On the other hand, LEDs are more rugged, less sensitive to temperature variations, and probably have a longer life.

As a result of these differences, LEDs are used in short single-hop path lengths, while lasers are used for long paths and multihop systems.

Although the spectral bandwidth of laser emissions is comparatively narrow, it is not monochromatic, and its fine-grained spectrum shows a number of discrete spectral lines. Figure 12.9 shows a representative example of the spectrum of a laser emitter at different power levels.

With a spectral distribution of this width, the bandwidth is reduced as the result of chromatic dispersion (see Sec. 12.4.3), and this can be a significant problem on long-distance high-speed circuits. For these circuits, the output spectrum can be limited to a single spectral line, often 1.55 μm, by the use of an external cavity or a resonant structure within the laser. The result is a *single longitudinal mode* or SLM laser.

12.8 Receivers for Fiber Optic Systems

A fiber optic receiver consists of a *photodetector* that converts an optical to an electrical signal, followed by stages of electronic amplification. The critical element in the receiver is the photodetector, since it determines the receiver's sensitivity.

The sensitivity of a fiber optic receiver is defined as the minimum received power P_r required to achieve an error probability P_e at a bit rate B. P_r can be calculated from the minimum number of *photons* n_p that must be received per pulse:

$$P_r = \frac{n_p Bhc}{\lambda} \tag{12.3}$$

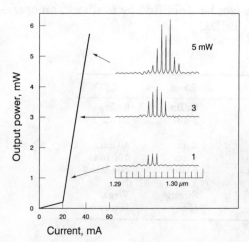

Figure 12.9 Laser emitter spectrum. (*From Ira Jacobs, "Fiber Optic Transmission Systems," chap. 8 in A. F. Inglis (ed.),* Electronic Communications Handbook, *McGraw-Hill, New York, Fig. 8.7.*)

where λ = wavelength
  B = bandwidth
  h = Planck's constant = 6.6252×10^{-34} J $\cdot$ s
  c = velocity of light = 3×10^8 m/s

n_p is calculated from the error probability: $P_e = \frac{1}{2} \exp(-n_p)$.
The most common photodetectors are the *avalanche* photodetector, APD, and the *PIN junction*. Their sensitivity in terms of the number of photons per bit and the received power level for a bit error probability of 10^{-9} for PIN and APD photodiodes is shown in Fig. 12.9.

The demands on the performance of a fiber optic receiver are severe. As an example, consider an APD receiver operating at a bit rate of 100 Mbits/s. It would require 10^2 photons per bit or 10^{10} photons per second to achieve a bit error probability of 10^{-9}. If each photon were converted to an electron by the photodetector (an optimistic assumption), the current would be 1.6×10^{-9} A or 1.6 nA. (This compares with an output signal of more than 100 nA for a typical photoconductive image tube in broadcast cameras.) A current of this level requires a high degree of amplification in the electronic system following the photomultiplier.

12.9 Passive Components

A number of passive devices, including connectors, optical attenuators, couplers, and wavelength-division-multiplexing (WDM) filters, some necessary and some optional, are available to complement the active devices and complete the system.

12.9.1 Connectors

The availability of connectors that could place and maintain the ends of very tiny fibers in almost perfect alignment at joints has been a challenging engineering assignment. The success of single-mode fiber was made possible by the development of tools and techniques for splicing extremely thin fibers with precision. Permanent connections are bonded, while demountable connectors utilize cylindrical bushings or conical plugs. Techniques have been perfected that have reduced the losses at splices to 0.05 dB or less.

12.9.2 Optical attenuators

Because of the wide amplitude range of received signals, a means of adjusting the optical input power is required for the operation of practical systems. This is accomplished by attenuators, both fixed and adjustable neutral-density filters.

12.9.3   Couplers

Couplers divide the power in a single fiber between two or more paths. This can be accomplished with a Y junction or, if only a small amount of power is desired in one of the branches, by putting a sharp bend in the fiber so that reflection is less than total and the power that escapes is captured by a second fiber.

12.9.4   Wavelength-division-multiplexing filters

The purpose of wavelength-division-multiplexing filters is to separate the radiation from the optical source into radiation bands, each of which forms a signal channel that is modulated separately. The channels are multiplexed and transmitted by a single fiber. This is a very effective technique for utilizing the large bandwidth of fibers to increase their information handling capacity.

The filters are thin-film dielectric materials that transmit one wavelength and reflect the other—as with the dichroic surfaces used in color television cameras.

References

1. Inglis, Andrew F., *Electronic Communications Handbook*, Figure 17.19, McGraw-Hill, New York, 1988.

13

Color Receivers

13.1 Introduction

The performance of today's television color receivers is superior by orders of magnitude to that of those available in 1953 when color television broadcasting was first approved. They produce far better images in every measurable respect (see Chap. 3), they make larger pictures with smaller sets, and they are far more stable and reliable. Best of all, the improved performance has been achieved while offsetting manyfold inflationary cost increases, and the average 1992 receiver is lower priced than in 1954, even in current dollars. The engineering resources of many countries have contributed to this progress, and it has truly been a superb worldwide achievement.

A remarkable feature of this performance is that it has been achieved without any revolutionary technical breakthroughs. The same superheterodyne principle continues to be used for the radiofrequency section of the receiver. The original three-gun shadow-mask principle continues to be used in most receivers. The transmission format established by the FCC in 1953 (and by the EBU in Europe) is essentially unchanged. As is often the case in war and technology, success lies in the details.

13.2 Receiver Configuration

A functional diagram of a typical television receiver is shown in Fig. 13.1. Its layout is conventional—a tunable RF stage for channel selection including a local oscillator and first detector, and an IF amplifier and second detector for video and audio. The video channel employs a diode detector or synchronous detector. The audio channel employs a discriminator or other FM detector. The output of the video channel includes the signal multiplexed with the timing and color synchroniz-

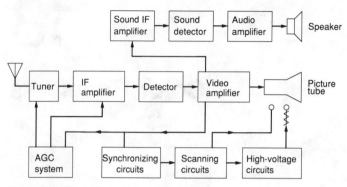

Figure 13.1   Typical receiver configuration.

ing signals in the format described in Sec. 3.14. The timing and color synchronizing signals control the kinescope sweep circuits and the color phase reference.

13.3   RF Amplifier and Channel Selector

Figure 13.2 is a diagram of an RF amplifier and channel selector.

VHF signals are amplified before mixing with the output of the local oscillator, which operates 45.75 MHz above the picture carrier and 41.25 MHz above the sound carrier. The level of the incoming carrier signal can vary from 10 μV to 100 mV, and the AGC amplifier maintains the signal level within reasonable limits. The maximum gain of the channel selector amplifier is approximately 20 dB.

Broadcast receivers can be tuned to the VHF and UHF channels in the television broadcast bands, channels 2 to 13 and 14 to 83. For reception of the additional cable channels, the receiver is tuned to channel 3 or 4 and the cable channels are heterodyned to it by means of a *set-top converter.* Alternatively, *cable ready* receivers are capable of picking up the standard cable channels (see Chap. 10) directly without an external converter.

The difficulty in designing a tunable UHF amplifier of reasonable price and performance often leads to the design shown in Fig. 13.2, where the channel selector has a loss rather than a gain. Its IF output is fed to the mixer of the VHF section, which operates as an amplifier under these conditions.

In earlier sets, VHF tuning was accomplished by variable capacitors with mechanical detents, while UHF tuning required the more difficult and time-consuming method of continuous tuning. The FCC then decreed that VHF and UHF receivers must have the same ease of tuning, and this requirement has been met by the use of electronically

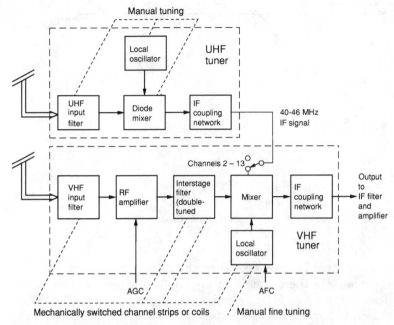

Figure 13.2   VHF and UHF channel selectors. (*From K. Blair Benson,* Television Engineering Handbook, *McGraw-Hill, New York, 1986, Fig. 13.2.*)

controlled systems that can have preset values for the desired channels. *Varactor tuners* that make use of the change in capacitance of *varicap diodes* are often used as the variable tuning elements in both the VHF and UHF local oscillators. The most advanced receiver designs employ a microprocessor that permits easy preset and readout of selected channels.

13.4   IF Amplifier

13.4.1   Frequency response—video channel

The frequency response of the video channel of the IF amplifier must meet rather exacting requirements; see Fig. 13.3. It must provide high rejection of the adjacent-channel picture carrier at 39.75 MHz and the adjacent-channel sound carrier at 47.25 MHz. The treatment of the channel sound carrier differs from system to system, but it must be highly attenuated at the input of the video detector. The picture carrier at 45.75 MHz should be located at the 50 percent (– 6 dB) point on the upper slope of the response curve. This position equalizes the sum of the responses of the sidebands (at 100 percent in Fig. 13.3) in the vestigial sideband area for baseband frequencies from 0 to 1.5 MHz. For modulating frequencies above 1.5 MHz, equalization of the

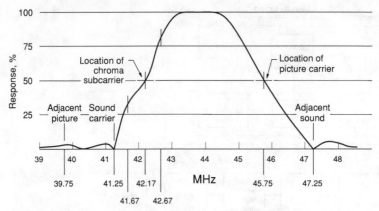

Figure 13.3 IF frequency response. (*From K. Blair Benson,* Television Engineering Handbook, *McGraw-Hill, New York, 1986, Fig. 13.17.*)

video channel must be provided when the response of the IF falls below 100 percent. In the example in Fig. 13.3, this occurs at sideband frequencies below 43.25 MHz or baseband frequencies above 2.5 MHz. Equalization is particularly important in the vicinity of the chroma subcarrier in the range 41.67 to 42.47 MHz. Figure 13.4 shows the details of the equalization required.

The presence of the subcarrier imposes requirements on the performance of the IF amplifier in a color receiver that are not present with monochrome. The bandwidth must be greater to accommodate the subcarrier and its sidebands. The tuning must be stable so that the correct ratio of luminance to chrominance signals is maintained. Linearity must be sufficient to avoid intermodulation between subcarrier and carrier. The envelope delay as well as the frequency response must be equalized over the channel passband.

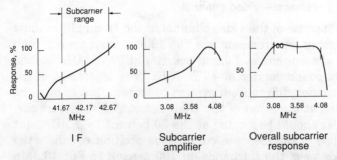

Figure 13.4 Color subcarrier equalization. (*From K. Blair Benson,* Television Engineering Handbook, *McGraw-Hill, New York, 1986, Fig. 13.18.*)

The frequency response is often established by a SAW (surface acoustic wave) filter; see Sec. 9.8.1.6.

13.4.2 Sound IF channel

The sound IF subcarrier operates at 41.25 MHz, 4.5 MHz below the video. It can be handled in a number of configurations. In one, it is carried through the IF stages at a level 10 to 20 dB below the picture carrier. The 4.5-MHz difference signal, including its frequency modulation, then appears at the output of the detector. In another, the sound IF is separated from the video at the output of the second IF stage and is then processed by a separate sound IF amplifier.

In both cases, the final difference between the picture and sound intermediate frequencies is 4.5 MHz as established at the transmitter, and they are known as *intercarrier sound systems*.

13.4.3 Gain

The IF amplifier gain is typically 90 dB, which is offset by 20 dB loss in the input and detector stages:

Stage	Gain, dB
Input coupling	-10
First IF stage	+30
Second IF stage	+30
Third IF stage	+30
Detector	-10
	+70

13.4.4 AGC and AFC

Automatic gain control (AGC) and automatic frequency control are standard features of most receivers.

Automatic gain control is often provided in both the RF and IF stages of the receiver. For low- and medium-level signals, gain control is applied only in the IF section. Additional gain reduction is applied in the RF section for very strong signals.

Automatic frequency control is applied to the local oscillator and maintains the IF carrier frequency at 45.75 MHz.

13.5 Detection

Detectors can be either diode or the more advanced synchronous and balanced type. The latter is more immune to cross-modulation between the video carrier, the color subcarrier, and the sound carrier. Overmodulation of the video carrier can result in an intercarrier beat signal and a buzz in the sound system.

13.6 Sync Separation and Scanning Generation

The vertical and horizontal sync pulses are transmitted at a higher modulation level than the video signal and are easily removed for driving the scanning generators.

The vertical scanning generator operates at the relatively low frequency of 60 Hz and is usually a multivibrator-type sawtooth generator that is synchronized by the vertical sync pulses as modified by the equalizing pulses.

The horizontal scanning generator operates at the much higher line rate, and the retrace time is much shorter. As a result, the voltage, $L(di/dt)$, developed across the deflection transformer and/or deflection yokes is very high. This voltage can be rectified and used to provide the anode voltage for the kinescope.

13.7 Subcarrier Demodulation and Video Amplification

The color subcarrier is removed from the composite video signal following the second detector and is demodulated to recover the I and Q signals (see Sec. 3.12.2). The I and Q signals are matrixed with the Y luminance signals to generate first the $R - Y$ and $B - Y$ and then the R, G, and B primary signals that are applied to the three kinescope electron guns. The video amplifiers at the output of the matrix circuits must be capable of amplifying the level of the video signal from 1 or 2 V to 20 V or more. Processing circuits for image enhancements may be added at this point.

The level of the subcarrier is adjustable at the input of the demodulator by the "color" control to establish the saturation of the image.

The phase of the subcarrier with respect to the color burst can be adjusted to control the hue, and this is the "hue" or "tint" (sometimes confusingly called the "color") control.

13.8 Color Display Devices

The display device, the component that forms the visual image, is both the end point and the key to the performance of color television systems. No matter how perfect the performance of the signal generating and transmission equipment, the image quality can be no better than that of the display device. Improvements in display devices, probably more than in any other single component, have been the greatest contributor to the dramatic advances in television receiver performance described in Chap. 2 and at the beginning of this chapter.

13.8.1 Shadow-mask kinescopes

Shadow-mask kinescopes were used as the display devices in the original color television receivers, and this principle has been remarkably durable and is still in wide use 40 years later. It is produced in a number of formats, two of which are shown in Figs. 13.5 and 13.6.

The format in Fig. 13.5 has round holes in the shadow mask arranged in triads and a "delta gun" structure with the electron guns arranged in a triangle.

The format in Fig. 13.6 has an "in-line" gun structure, with the guns mounted side by side rather than in a triangle and with slots rather than round holes in the shadow mask. An advantage of this configuration is that there is no color registration problem in the vertical direction.

As a result of the relative positions of the electron guns, shadow-mask holes, and tricolor phosphors, the electrons from each gun impinge only on phosphors of one color. If the amplitude of the signal current from each gun is controlled in accordance with the brightness of the primary component of the scene at each point, the amplitude of the image brightness for each primary color will correspond to the scene brightness for that color.

A black deposit around each phosphor dot (the surround) is an es-

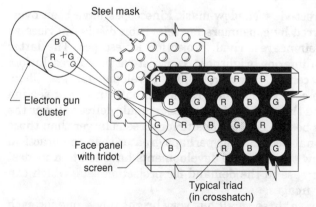

Figure 13.5 Shadow-mask kinescope, delta gun, round hole. (*From K. Blair Benson*, Television Engineering Handbook, *McGraw-Hill, New York, 1986, Fig. 12.4.*)

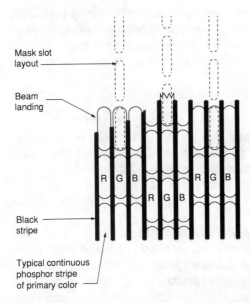

Mask slot
layout

Beam
landing

R G B   R G B
       R G B

Black
stripe

Typical continuous
phosphor stripe
of primary color

Figure 13.6 Shadow-mask kine-scope, in-line gun, stripes.

pecially important feature of the shadow-mask design. It absorbs the ambient light striking the faceplate so that blacks appear to be really black, thus increasing the contrast ratio. See Sec. 2.15.2.

In addition to the formats shown in Figs. 13.5 and 13.6, the shadow-mask principle is applicable to a number of additional formats that vary the shape and location of the holes in the masks and the arrangement of the guns.

13.8.2 Projection kinescopes

Until recently, direct-view shadow-mask kinescopes have been overwhelmingly preferred by consumers for television display devices as compared with their nearest rival, projection kinescopes. The latter have suffered from inferior brightness, contrast ratio, and resolution, and high costs. Recently, however, HDTV developments (see Chap. 14) have been the catalyst for improvements, both in the projection sets themselves and the quality of broadcast signals delivered over the air. There also has been a growing desire for screens larger than those with a 36-in diagonal—equal to or perhaps greater than the practical limit for direct-view tubes. These developments have led to a modest but significant increase in the demand for projection sets, which can produce far larger images.

Projection sets have three small but very bright tubes, one for each primary color. The images on the tubes are enlarged, converged, and focused on a screen by means of an optical system that may employ mirrors, lenses, or both. Impressive improvement has been made in their performance in recent years, although it is still somewhat marginal with respect to brightness and, for their very low viewing ratios, their resolution. See Chap. 14 for further discussion.

13.8.3 Other display devices

A number of other display devices have been developed, most of them with limited success. Two, the eidophor and luminescent panels, will be mentioned here.

The *eidophor* is a projection system developed in Switzerland for theater television and other applications requiring a large and bright display. Unlike conventional projection systems, in which the total light available is limited by the output of a cathode-ray tube, the primary source for the eidophor is an arc light or equally bright source. The image is formed by a *Schlieren optical system,*[1] a complex electro-optical system.

Luminescent panels are presently under development, and hold great promise for future display devices. A flat luminescent surface is divided into a multitude of individual cells, which become the pixels. Each cell emits light as the result of electrical or optical excitation. The design and manufacturing problems are formidable at the present time, and this must be classified as a future development.

13.9 Receiver Performance Criteria

The performance criteria of television receivers can be divided into two categories: the performance of the signal amplification sections and the performance of the image display device.

The primary criteria of the signal amplification sections are the noise figure and the selectivity. These determine the receiver's performance with weak signals and in the presence of adjacent-channel interference. The primary criteria of the display device, in addition to its size, are its resolution (including the convergence of the three primary colors), its brightness, and its contrast ratio.

13.9.1 Noise figure

The noise figure criterion was originally developed for specifying the performance of radar receivers. It is defined as the signal-to-noise ratio at the output of a receiver compared to the signal-to-noise ratio at its input when the input is at room temperature T_0 (assumed to be 290 K or 17°C).

Since the receiver inevitably adds noise, the noise figure expressed as a fraction will be less than unity, and expressed in decibels will be negative. To avoid this, it is conventional to express the noise figure in decibels as a positive number as shown in Eq. (13.1).

$$N_f = -10 \log \frac{(S/N)_{output}}{(S/N)_{input}} \tag{13.1}$$

(when $\qquad T_{input} = T_o = 290$ K)

The receiver's noise figure is a critical performance criterion for receivers located in areas of weak local signals.

Typical noise figures are 4 to 8 dB on VHF channels and 7 to 12 dB on UHF channels.

13.9.2 Selectivity

The selectivity of a receiver is a measure of its ability to reject off-channel signals. Interfering off-channel signals can be adjacent channels that are not sufficiently attenuated by the selectivity of the receiver or spurious signals such as harmonics or heterodyned cross-modulation products.

The receivers's resistance to adjacent-channel signals is largely determined by the response of the IF system; see Fig. 13.3. Interference from cross-modulation products is minimized by the FCC's allocation plan, particularly for the UHF channels, that avoids the assignment of image frequencies and other potential off-channel sources of interference in the same area. See Sec. 9.16.

Reference

1. W. Good, "Projection Television," *IEEE Trans.*, CE-21(3), August 1975.

HDTV Systems

14.1 Introduction

HDTV, or *high-definition television,* was initially a technical term with a narrow meaning describing television images with sharp transitions between their dark and light areas so that they had the appearance of being in good focus (see Chap. 2). As the term passed into the vernacular, its meaning was broadened, and it is now used to describe images that are of exceptional quality in all respects, including their signal-to-noise ratio, gray scale, colorimetry, and freedom from spurious signals. Its meaning has been broadened still further to describe the systems of equipment that generate and display the images as well as the images themselves. Finally, it has become descriptive of an entire communications *medium.*

The broadening of the meaning of HDTV requires a consideration of sound quality, since the visual image is only part of a total audiovisual experience. HDTV systems must include high-fidelity audio facilities.

14.2 HDTV Performance Objective

The ultimate performance objective of HDTV systems is to generate image quality that equals or exceeds the ability of the human eye to perceive it—quality that is sufficiently high that further improvement would be unnoticeable. The quality level at which this occurs depends on the viewing ratio—the lower the ratio, the more stringent the requirements on the system (Secs. 2.7.4 and 2.16.3).

The performance objectives can also be expressed by comparison with motion-picture film. While NTSC broadcast television has approximately the same performance as 16-mm film, HDTV performance should approximate that of 35-mm.

14.3 HDTV Performance Parameters

14.3.1 Ideal system performance

The performance parameters of an ideal HDTV system are shown in Table 14.1, an ideal system being defined as one that produces image quality that meets or exceeds the perception capability of the human eye at the minimum practical viewing ratio, usually specified as 2:1.

14.3.2 Systems with performance compromises

The enormous bandwidth and other costly requirements of systems meeting the ideal as shown in Table 14.1 make them impractical for most applications, even presuming that they are technically feasible. As a result, the development of HDTV systems is proceeding along three lines, each with a different degree of compromise with the ideal. They are:

1. Production systems designed for the generation of signals that produce images of a quality that approaches the ideal and that include the smallest compromises. Generally, the signals generated by these systems are not transmitted in their original format but are recorded on tape or film or transcoded to HDTV formats with lower bandwidth requirements.

2. Systems designed for the production of signals to be transmitted over satellites, which do not have the severe bandwidth and regulatory limitations of the standard broadcast channels.

3. Terrestrial broadcast systems designed to operate in the existing broadcast channels with limited bandwidth, 6 MHz in the United States. These systems have the greatest performance compromises, and they present the greatest technical challenges. Because of the potential economic importance of broadcast systems, much of the HDTV engineering development work now in progress is devoted to systems that can produce the highest possible signal quality within the limits of the existing broadcast channels.

14.3.3 Production system performance

After extensive discussion, negotiation, and study by engineering groups in all countries with extensive television technical resources,

TABLE 14.1 Television System Comparison

	Broadcast NTSC		HDTV	
	Low-cost	Standard	Ideal	SMPTE 240M
Aspect ratio	4:3	4:3	16:9	16:9
Minimum viewing ratio	6:1	5:1	2:1	2.5:1
Signal format	Composite	Composite	Component	Component
Transmission format	Analog	Analog	Digital	Digital
Compatible	Yes	Yes	No	No
Scanning format	Interlaced	Interlaced	Sequential	Interlaced
Number of scanning lines	525	525	1125	1125
Field rate, fps	59.94	59.94	60	60
Bandwidth, MHz (uncompressed)				
Luminance	3.2	4.2	60	30
Chrominance	0.5–1.5	0.5–1.5	30–30	15–15
Limiting Resolution—Luminance Signal (TV lines)				
Horizontal	253	332	888	882
Vertical				
Static	137	276	661	661
Motion	137	137	661	331
Horizontal aperture correction	Partial	Yes	Yes	Yes
Vertical aperture correction	Partial	Yes	Yes	Yes
Horizontal image enhancement	No	Yes	Yes	Yes
Definition (TV lines)				
N_{ehoriz}	210	290	550	550
N_{evert} (static)	93	200	500	500
N_{evert} (motion)	93	100	500	300
Aliasing				
Minimum vertical image detail for scanning line aliasing (TV lines)				
Static	>196	196	>472	>472
Motion	>98	>98	>472	>236
Artifacts				
Interline flicker	Yes	Yes	No	Minor
Cross-color	Yes	Yes	No	No
Signal/noise (unweighted), dB	>35	>45	>52	>50
Gray scale				
Brightness, foot-lamberts	40	65	100	50
Contrast ratio, dB	20	65	100	50
Chromaticity coordinates*	SMPTE C	SMPTE C	Not applicable; three-primary system cannot reproduce full gamut of colors	SMPTE 240M
or alternatively	SMPTE C	NTSC		SMPTE 240M (Appendix)

*See Fig. 14.4.

the HDTV production standard SMPTE 240M, was adopted in 1987, the first HDTV standard to be adopted.

Table 14.1 compares the performance of a 240M system with that of an ideal system. To provide an additional frame of reference, the performance is also compared with that of two NTSC systems operating at different quality levels.

The specifications contained in the 240M standard are described in subsequent sections.

14.3.4 Satellite broadcast system performance

The bandwidth of the channels in satellite broadcasting systems, typically 25 or 40 MHz, is substantially greater than that of the existing 6-MHz terrestrial broadcast channels. This not only permits wider video basebands, but also provides a wide variety of options in the choice of modulation formats.

The first HDTV broadcasting system was operated by NHK in Japan and employed satellites. Its capability demonstrated the potential of DBS (see Chap. 11) for HDTV broadcasting.

The NHK transmission format is known as MUSE, for multiple sub-Nyquist encoding and decoding (see Sec. 14.11.2). Its equivalent base bandwidth is 8.1 MHz, and it has 1125 scanning lines. See Sec. 14.11.2 for a more complete description of MUSE.

Satellite broadcasting requires high-power satellites with wide orbital spacing in order to permit the use of small receiving antennas with low gain and directivity.

14.3.5 Terrestrial broadcast system performance

Terrestrial broadcast HDTV signals must fit into the 6-MHz broadcast channels now occupied by signals in the NTSC format. Substantial bandwidth compression is required for satisfactory performance of HDTV systems using these comparatively narrow 6-MHz signal channels, and to a large extent the quality of the images produced by these systems depends on the effectiveness of the bandwidth compression technology.

Rapid improvements are currently being made in bandwidth compression systems, both analog (Sec. 14.9) and digital (Sec. 4.18). As of this writing (1992), digital technologies are gaining favor and will probably be adopted.

The steady improvements in bandwidth compression technologies create a moving performance target, but even at present, the quality of 6-MHz HDTV images perceived at low viewing ratios is substantially superior to that of NTSC images.

14.3.6 HDTV, EDTV, IDTV, and ATV

The performance gap between NTSC and HDTV systems, even HDTV systems that operate within a 6-MHz transmission bandwidth, suggests the need for systems of intermediate performance. These systems are generally described as extended or enhanced definition (EDTV) or improved definition (IDTV). The meanings of these terms are not precise, and the dividing line between HDTV and EDTV is not sharp. IDTV has come to mean primarily systems with improved receivers. HDTV, EDTV, and IDTV systems are known collectively as advanced television (ATV) or advanced definition television (ADTV).[1]

14.4 Compatibility—Definition and Status

Compatibility is an issue primarily where HDTV signals share broadcast channels with NTSC signals. For complete compatibility, an NTSC receiver tuned to an HDTV signal must display a normal NTSC image, including adjustment to a different aspect ratio, a different number of scanning lines, and possibly different field and frame rates.

Differences in aspect ratio are inherently incompatible, and the viewer of an NTSC receiver tuned to an HDTV channel must be satisfied with the center portion of the 16:9 image.

With digital signal processing, field and line rates can usually be converted or *transcoded* to other rates, so that systems with different rates can have a degree of compatibility. If the conversion is to a lower rate, some information will be lost. If the conversion is to a higher rate, new information must be generated by interpolation or prediction. Conversion systems are complex and costly and probably beyond the range of consumer products at this time.

The problems of compatibility are increased when they are combined with the necessity for bandwidth compression. As a result, a number of proposals were made that would combine an NTSC-compatible channel with an *augmentation channel* that would carry the additional signal information required for HDTV-quality performance. This option, however, was foreclosed by an FCC policy statement in 1990.

14.5 FCC HDTV/EDTV Compatibility Policy

To date, no practical system has been proposed that meets the tests of HDTV performance and compatibility, at a reasonable cost. This was recognized, explicitly or implicitly, in a policy statement with respect to HDTV/EDTV standards and channel assignments for *broadcast* stations announced by the FCC on March 21, 1990. This policy does *not* apply to satellite broadcasting or to program production, but it considerably narrows the choices for broadcast systems.

The major provisions of the FCC HDTV policy statement are as follows:

1. HDTV/EDTV channels will be separate from NTSC channels, but will have the same 6-MHz bandwidth. The result will be two classes of channels—NTSC and HDTV. It is probable that the HDTV channels will require digital transmission with major bandwidth compression to achieve HDTV performance.

2. The use of separate "augmentation channels" to carry the higher-frequency components is not encouraged.

3. Compatibility with NTSC, except for the 6-MHz channel bandwidth, is neither required nor expected of the signals in the HDTV channels.

4. Simulcasting, broadcasting the same program material on separate channels with NTSC and HDTV/EDTV standards, is encouraged. This would provide a mechanism for an orderly transition from NTSC to HDTV.

   Simulcasting has the serious problem that there is no vacant spectrum space for the new simulcasting channels in the most populous parts of the country under current assignment policies. Digital transmission offers the hope of solving this problem, both by allowing closer spacing between co-channel stations and by the use of "taboo" channel assignments.

   Compatibility between HDTV/EDTV signals on simulcast channels will be required, and a single standard will be adopted for these channels.

5. EDTV standards will not be selected until an HDTV standard has been chosen. (To do so would jeopardize the development of HDTV because of the public's prior investment in EDTV receivers.)

These conditions, which were based on the FCC's confidence that technology would provide a satisfactory method of transmitting an HDTV signal in a 6-MHz channel once the compatibility requirement was removed, has had a major effect on the prospects of the candidates for selection as the broadcast standard. It also affected the priority of the engineering programs. It eliminated all broadcast systems that required more than a 6-MHz bandwidth, and, except for channel bandwidth, NTSC compatibility ceased to be required. It improved the prospects for the selection of a digital standard because the digital format has the most promise for reducing channel bandwidth and the spacing between co-channel stations—as would be required to make room for simulcast channels. Further, compatibility with analog transmissions is no longer an issue.

As a result of this policy, prospective HDTV/EDTV systems in the United States can be divided into two classes:

1. Systems employing 6-MHz transmission channels for simulcast broadcasting.

2. Systems employing wider-band channels that would be used for satellite program distribution and nonbroadcast applications.

14.6 HDTV Technical Challenges

Three basic technical challenges were faced in the engineering development of systems for the generation, recording, transmission, and display of HDTV signals for use as a broadcasting medium.

The first is the incompatibility of HDTV signals with existing broadcast standards. This problem did not have a completely satisfactory solution, and the latest FCC policy statement avoids it by permitting the simulcasting of an incompatible HDTV channel (see Sec. 14.5).

The second is the lack of space in the television broadcast spectrum to accommodate the enormous bandwidth requirements of HDTV signals. This problem appears to have a satisfactory, although not perfect, solution through the use of bandwidth compression technology.

The third is equipment performance, particularly of cameras, recorders, and display devices (see Chaps. 5 and 6). These barriers are also lowering rapidly with the introduction of new technologies, and as of this writing, the greatest remaining challenge is the development of a large, bright, flat-screen display device that can be manufactured at a reasonable cost.

14.7 HDTV Performance Parameters

The performance parameters for the various classes of HDTV systems were summarized in Sec. 14.3 and Table 14.1. The following sections describe these parameters in more detail.

14.7.1 Aspect and minimum viewing ratios

The aspect and minimum viewing ratios head the list in Table 14.1 because they define the essence of the differences between HDTV and standard broadcast TV.

Consistent with the trend to "wide screen" in motion picture images, almost all HDTV systems utilize an aspect ratio of 16:9 (see Sec. 1.5) rather than the 4:3 ratio employed in broadcast television. The wider screen, together with a shorter viewing distance (and higher-quality audio), gives HDTV images a sense of "presence" that cannot be achieved with broadcast standards.

As described in Chap. 2, the perceived sharpness of an image is determined by the limitations of the human eye as well as by the image's geometric properties. At distances greater than about 5 times picture height, a viewing ratio of 5:1, the normal eye is unable to perceive a significant difference in the sharpness of HDTV and NTSC broadcast images. The benefits of HDTV transmissions, therefore, can only be obtained with viewing ratios that are significantly less than 5:1. In most viewing situations this requires a considerably larger screen than is used in typical receivers. *Large-screen TV*, therefore, is perhaps a more meaningful description than *HDTV*.

The minimum viewing ratios shown in Table 14.1 are only approximate, since it is not a precise specification. At viewing distances less than the minimum, the eye will be aware of the lack of perfect picture sharpness. Noise, aliasing, and other image imperfections may also be visible. At greater viewing distances, the image will contain fine detail that the eye is unable to perceive.

The larger minimum viewing ratio for the low-cost NTSC systems as compared with their standard NTSC counterparts reflects their slightly poorer performance.

14.7.2 Video signal formats

As described in Chap. 3, video signal formats for color transmission can be either *composite* or *component*. Composite formats require less bandwidth and are used for broadcasting, but they have certain inherent defects, such as cross-color and cross-luminance artifacts, that are not present in component systems. The bandwidth penalty for the component format is so great, however, that it is not used for analog transmissions. It is used with digital transmissions (often with bandwidth compression), for certain editing operations where multiple-generation recordings are required, in some closed-circuit applications in some wider recording formats, and in satellite broadcasting. It is frequently combined with the MAC format (see Sec. 3.20).

Most HDTV systems that are proposed for use in applications that have no severe bandwidth limitations or compatibility requirement employ the component format. This eliminates or reduces several of the problems of composite systems that become significant in HDTV systems.

14.7.3 Transmission formats

Analog transmission was almost universally used during the early years of HDTV, but this may change in the future. The bandwidth reduction potential of digital transmission, together with its lower level of co-channel interference, create the probability that HDTV broadcast signals will be digital.

14.7.4 Interlaced and sequential scanning

The tradeoffs in the choice between interlaced and sequential scanning and the number of scanning lines were described in Sec. 1.4. The priorities in making these selections are somewhat different in broadcast and HDTV systems because aliasing and other scanning-related defects are a more serious problem for HDTV. They can be minimized by increasing the number of scanning lines and by the use of sequential scanning. For moving objects, the effective number of scanning lines is only one-half the total, and the ideal system in Table 14.1, therefore, employs sequential scanning.

In addition to these inherent defects, interlaced scanning has the practical defect that adjacent lines tend to pair, i.e., to fail to achieve interlace owing to small timing errors in the receiver. This is assumed in the low-cost system in Table 14.1, where the number of scanning lines is effectively reduced by one-half, with a corresponding negative effect on $N_{e,\text{vert}}$, vertical resolution, and aliasing.

Even with normal interlace, images of objects in motion have artifacts and other spurious effects because of differences in adjacent lines as the object moves.

Note the differences in the performance of the ideal and 240M systems with respect to moving objects and artifacts owing to the use of interlaced scanning with the latter.

In spite of these defects, the bandwidth penalty of sequential scanning is so great that most HDTV systems continue to employ interlace. This is true even of systems employing the very demanding 240M standard. Sequential scanning may become more widely used in the future if digital bandwidth compression systems are shown to be sufficiently effective.

14.7.5 Number of scanning lines

Increasing the number of scanning lines increases the vertical resolution and $N_{e,\text{vert}}$ while reducing aliasing and line visibility. The latter parameters are as important as resolution and $N_{e,\text{vert}}$ in determining the performance of HDTV systems and are given greater weight in the choice of the number of scanning lines than in NTSC systems.

The number of scanning lines chosen for most HDTV systems is approximately double the 525 lines of NTSC systems. Industry agreement has not been reached on the exact number, but three values— 1125 lines, as shown on Table 14.1, 1050 lines (or 1049 for interlaced systems), and 1250 lines—have been actively proposed. The more common choice is 1125 lines. It is used by NHK for its MUSE (multiple sub-Nyquist sampling and encoding) system, the first HDTV system to be used for broadcasting to the public (via satellite), and is specified by the SMPTE for its HDTV production standard, 240M. The Sarnoff Laboratories proposed 1050 lines for its ACTV system. Since 1050 lines is double the standard 525 rate, it has the advantage that it can easily be transcoded to and from a 525-line signal and used as a means of bandwidth reduction by interpolation (see Sec. 14.8.3). Also, 1250 lines has been proposed as an international standard.

The angular separation of the scanning lines in a 1125-line system at a viewing ratio of 2 is 1.8′, and with that separation, they are barely visible as separate lines. Aliasing begins for vertical detail components with line numbers greater than 950 (the number of active scanning lines). The amount of signal energy generated at these line numbers is extremely low, even with the best cameras, and aliasing is minimal.

Unlike NTSC systems, where the number of scanning lines is constant throughout the system, the number of scanning lines in an HDTV system may be transcoded from camera to transmission system to display device. For example, an image may be scanned at 1050 lines, transcoded to 525 for transmission, and transcoded back to 1050 lines for display. Some information will be lost at each transcoding, but the definition and aliasing are improved as compared with an all-525 line system, and the transmission bandwidth requirements are no greater.

14.7.6 Field rate

It would be desirable to establish worldwide scanning standards, including the field rate, for HDTV systems, and serious international efforts have been made to reach agreement. To date, the difference between the 50- and 60-fps field rates, which are employed in different countries for their broadcast systems, has posed a seemingly insurmountable problem for achieving this objective.

All major countries now propose the same field rate for composite HDTV signals as for their broadcast systems. The field rate for component systems is an integral number rather than a submultiple of the subcarrier frequency, e.g., 60 Hz rather than 59.94 Hz.

Unfortunately, it is difficult and costly to transcode signals from 60 fps to 59.94 fps, and these rates are basically incompatible for use in broadcast systems.

14.7.7 Bandwidth

The bandwidth of HDTV systems for broadcasting is their most critical parameter, and one of the most difficult to achieve for the broadcast service because of limitations on spectrum space. The bandwidths shown in Table 14.1 are for the basebands and assume no bandwidth compression. They are lower for the chrominance signal components, which is consistent with the reduced acuity of the human eye for the red and blue primaries.

The uncompressed bandwidth requirements of systems operating in accordance with the SMPTE Standard 240M—one 30-MHz channel and two 15-MHz channels (SM 240M does not specify bandwidth, but these bandwidths are required for consistency with the scanning standards)—can be met only in nonbroadcast systems.

Nearly all HDTV systems employ some form of bandwidth compression, and nearly equivalent performance can be achieved in narrower bandwidths with new compression technologies, particularly digital formats.

14.7.8 Limiting resolution

The vertical and horizontal resolutions in Table 14.1 were calculated by means of Eqs. (2.1) and (2.2).

The horizontal resolution achievable with the 30-MHz (uncompressed) bandwidth specified for the 240M system exceeds the capability of the human eye, but it may be important in production systems with multiple generations. It produces square pixels, i.e., their horizontal dimension is approximately equal to the spacing between scanning lines. (Square pixels simplify the geometric manipulation of digitized signals.) The horizontal resolution is greater than the vertical because of the Kell factor.

Approximately equivalent (but not quite equal) performance can be achieved in a narrower band by the use of bandwidth compression techniques.

In the NTSC composite format, the chrominance bandwidths are established by the color difference components, I and Q. The bandwidth of these components is adjusted in accordance with the acuity of the eye. The bandwidth of the I component (orange, cyan) is 0.5 MHz, while the bandwidth of the Q component (green, magenta) is 1.5 MHz.

The limiting resolution of HDTV systems employing large projection sets will be somewhat less than the 240M standards, even with the greatly improved designs of recent sets. The limiting resolution of the best projection sets is currently about 700 lines. It can be expected, however, that this will continue to improve in the future.

14.7.9 Aperture correction and image enhancement

Aperture correction and image enhancement (Sec. 2.4) are standard features of high-quality television systems. They are applied in the camera, and the resulting image sharpening is available to receivers of all cost levels. Additional aperture correction and image enhancement are included in HDTV and higher-quality NTSC receivers. Correction in the receiver has the advantage that adjustments can be made while observing the image, thus avoiding overpeaking.

14.7.10 Image definition—N_e

The equivalent line numbers for N_e, as tabulated in Table 14.1, are the best single criterion of HDTV and EDTV image definition. The total varies widely even within the home broadcasting service, from 93 to 290 lines, and the variation is much higher for HDTV and EDTV.

N_e depends not only on the number of scanning lines and bandwidth, but also on the aperture response of the system components. Very high quality components would be required to achieve the N_e values shown in Table 14.1 for HDTV systems.

14.7.11 Scanning line aliasing

Scanning line aliasing (see Sec. 2.17.6.1) occurs when the number of scanning lines is less than twice the line number of the highest vertical frequency component in the image. The result is a moiré or beat pattern, often very annoying with images containing a large amount of fine detail. The line numbers shown in Table 14.1 are equal to the number of active scanning lines.

14.7.12 Artifacts

Artifacts are spurious signals created artificially (hence the term *artifact*) by the imaging process. One of the most common is cross-luminance, a characteristic of composite systems employing a color subcarrier. It is a dot pattern that results from failure of the subcarrier signals on successive frames to cancel each other completely, e.g., on vertical edges of areas with high saturation.

They can also be produced by moving objects in an interlaced scanning and appear as interline flicker.

They can be eliminated or greatly reduced by the use of sequential scanning and component color systems.

14.7.13 Signal-to-noise ratio

The signal-to-noise ratios in Table 14.1 are unweighted so that they can be compared directly. See Sec. 2.16 for description and discussion.

Ideally, HDTV systems should be designed so that their weighted S/N at a low viewing ratio, say 2.5, is higher than the weighted S/N for NTSC systems at viewing ratios of 4 to 5. This fulfills the desire that HDTV images appear less noisy, even when viewed close up.

The relative magnitude of the noise weighting factors for NTSC and HDTV signals is determined by two offsetting effects. The lower viewing ratio that must be assumed for HDTV systems leads to a lower factor. On the other hand, much of the energy in HDTV signals is concentrated at higher frequencies, and this leads to a larger factor.

14.7.14 Brightness and contrast ratio

The combination of requirements for HDTV display devices—a large screen, high definition, and high brightness—creates design problems that have not been totally solved to date.

On a large screen, the available light is spread over a larger area and the brightness is lowered. Similarly, most design choices that are made to improve the definition, for example, reducing the size of the apertures in a shadow-mask kinescope, reduce its brightness. The problem is even more severe in projection-type display devices that are often used to achieve the desired size. The result is that, even with major improvements that have been made in recent years, currently available display devices fail to achieve the desired brightness and contrast ratio by a ratio of 2:1 or more; see Table 14.1. This means that better results will be achieved with current large-screen display devices when the picture is viewed in semidarkness.

14.7.15 Colorimetry

An examination of the CIE Chromaticity Diagram shows that it is impossible to achieve perfect matching of all the colors in nature with a three-primary system. No triangle on the diagram can encompass all the hues in the visible spectrum, but HDTV systems should seek to extend the gamut of the SMPTE C primaries that are commonly used in standard broadcast systems. The triangles in Fig. 14.1 show the sets of color primaries that have been standardized by the FCC and SMPTE for NTSC broadcast and HDTV systems. The NTSC primaries were proposed to the FCC and adopted in 1953 (Sec. 3.9.1). The

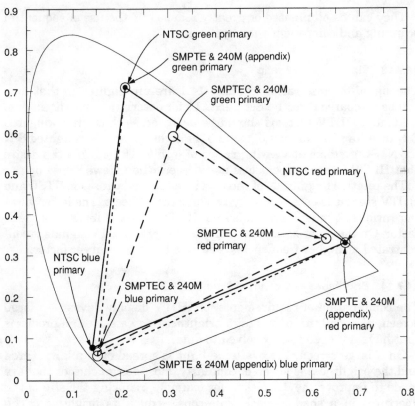

Figure 14.1   SMPTE 240M primary coordinates.

SMPTE C primaries were approved in 1982. They did not include as large a gamut of colors as the NTSC primaries, but they permitted much greater screen brightness with available phosphors (Sec. 3.9.2). In spite of their limited gamut, they were also incorporated in the SMPTE 240M standards for HDTV production. With the expectation that screen brightness would not be forever limited by existing phosphors, however, the SMPTE issued an appendix to its 240M standard with primaries having a color gamut that was the same as that of the original NTSC primaries except for a slight change in the blue.

As with the transfer characteristic, the colorimetry of film must be considered in specifying the colorimetry of an HDTV display intended for film transfer.[2]

4.17.16   Gamma

The gamma of NTSC signals is based on the assumption that the gamma of the display device, usually a shadow-mask tube, has a fixed value of 2.2 approximately and the signal gamma is adjusted to produce an overall gamma of unity. It is by no means certain that all HDTV display devices will have the same fixed gamma, and one of the problems of HDTV system designers is to develop a means for solving the problem of display devices with different gammas. One suggestion is to fix the gamma of the camera chain at unity and make the necessary gamma adjustments for the display device at the receiver.

When HDTV is used for program production and the objective is to transfer the end product to film, the transfer characteristic of film must be considered in planning the transfer characteristic of the television system.[3]

14.8 Bandwidth Reduction Technologies

The enormous bandwidth of unprocessed HDTV signals, which far exceeds that available in the broadcast spectrum, has led to intensive efforts to reduce their bandwidth with a minimum loss of image quality. With the motivation for compatibility diminished by the FCC policy decision, bandwidth reduction has become the focus of a major portion of the current HDTV engineering effort. The objective, simply stated, is to develop a technology that produces the highest possible image quality with a 6-MHz simulcast channel at a price consumers can afford.

In order to accommodate the 16:9 aspect ratio and to at least double the horizontal and vertical resolution, a minimum bandwidth reduction of 5:1 is required, i.e., a 6-MHz channel must transmit as much *useful* information as a 30-MHz channel with no bandwidth reduction.

The results of this effort have been quite successful, particularly with digital signals (see Chap. 4), owing in large part to the repetitive and predictable nature of video signals. Bandwidth reduction would not be possible if the signals were completely random and their instantaneous values were unrelated to previous or succeeding values.

In spite of the success of bandwidth reduction technologies, however, they always result in some loss of quality, albeit very small and often only with unusual image content.

Although the details of their technologies differ, most bandwidth reduction methods treat the signal components representing the stationary portions of the image differently from the dynamic components. The stationary portions do not need to be sent rapidly, which reduces their bandwidth requirements, but in the absence of special processing, dynamic components will suffer a loss of resolution or other defects with reduced bandwidth. This loss can be accepted, since the eye is not as sensitive to definition in moving objects, or the resolution can be partially restored by motion compensation, which is almost invariably used in bandwidth reduction systems.

Bandwidth compression can be accomplished in either the analog or digital mode. Digital signals are inherently wider band, but bandwidth compression can be more effective because of the ease and flexibility with which digital signals can be manipulated. As a result, bandwidth-compressed digital signals can or may have a lower bandwidth than analog.

Digital bandwidth reduction methods were described in Chap. 4. Representative analog bandwidth reduction technologies, including those that make use of sampling, are described in the following sections.

14.8.1 Interlaced scanning

Interlaced scanning, described in Sec. 1.4.1, was the earliest form of bandwidth reduction. It reduces the required bandwidth by a ratio of 2:1, although at the cost of some reduction in quality.

14.8.2 Frequency interlace

In the analog mode, frequency interlace is the horizontal counterpart of interlaced scanning. It achieves a nominal 2:1 bandwidth reduction by utilizing the gaps in the frequency spectrum of a video signal; see Sec. 1.11.1. Some of these gaps are occupied by the color subcarrier in the NTSC color system, but Fukinuki[4] noted that there were additional gaps that could be utilized by an augmentation subcarrier to transmit the high-frequency components of the signal.

As with the color subcarrier, it is desirable that the frequency of the augmentation subcarrier be an odd multiple of one-half the line-scan frequency.[5]

Also as with the color subcarrier, crosstalk into the luminance channel can be a problem with certain types of program material. It can be reduced by the use of a comb filter that separates the subcarrier and signal sidebands.

14.8.3 Subsampling and interpolation

The subsampling and interpolation technology described in Sec. 4.18.4 for digital signals can also be used for processing the samples of analog signals without subjecting them to digital encoding. This technology can be used not only to increase the horizontal resolution but also to double the number of scanning lines with no increase in bandwidth. For example, the signal can be transmitted in a standard 525-line format; at the receiver it is time-compressed and alternate lines are generated by averaging the samples from the lines above and below. The result is a 1050-line display.

14.8.4 Time compression integration (TCI)

TCI is a variation of the subsampling and interpolation technology and is the basis of the MUSE system developed and employed by NHK.[6]

The analog signal is sampled and stored, but only a fraction of the samples, e.g., the samples from every third field, are read out. In this example, a point on the raster receives a signal only once every three frames. This causes no problems for stationary objects, but the resolution of moving objects is reduced. This problem can be corrected by motion compensation circuits that interpolate the missing information (see Chap. 4).

14.9 Comparison of Analog and Digital Bandwidth Compression

A comparison of analog and digital bandwidth reduction must consider two processes, *source encoding* and *channel encoding*.

Source encoding is the processing of the digitized video baseband signal by one or more algorithms to achieve the desired bandwidth reduction. All proposed HDTV simulcast systems utilize digital source encoding to some extent, although other techniques such as scanning and frequency interlace may be used in addition.

Channel encoding determines the transmission format, including the choice of analog or digital, and analog and digital systems are defined by their method of channel encoding for transmission. Systems that employ both source and channel digital encoding are described as all-digital systems. Systems that combine digital source encoding with analog channel encoding are known as analog or hybrid systems.

The bandwidth reduction and motion compensation techniques for systems with digital channel encoding were described in Chap. 4. These methods are extremely powerful, and were it not for questions about the suitability of digital transmission in a broadcast environment, with its wide variations in signal levels and its potential for interference (see below), the adoption of digital channel coding as the standard for simulcast HDTV channels would be almost certain. Even with these problems, the probability is high.

Bandwidth reduction always results in some loss in picture quality because of the lower rate of information transfer. The loss can be small, however, and is largely confined to objects in motion. Some of this loss can be restored by motion compensation circuitry.

Prior to 1990, all proposals for the simulcast channels were hybrid or analog systems. In 1990 and 1991, four all-digital systems were proposed as the consequence of technical developments and the FCC policy statement that NTSC compatibility was no longer required (Sec. 14.5).

The digital format offers a number of important potential advantages for broadcasting:

1. It requires much less transmitter power and thus can probably be used to employ the UHF "taboo" channels without causing interference to existing analog signals.

2. It permits more powerful signal processing and thus has a greater potential for bandwidth reduction and motion compensation. This feature is also an advantage for broadcast-quality transmissions because it permits a number of channels, eight or more, to be multiplexed in the spectrum space occupied by a single conventional NTSC channel.

3. It provides a higher signal-to-noise ratio and more freedom from man-made interference, *provided the signal level exceeds a threshold value.*

4. It is more tolerant of interfering digital signals on the same channel.

5. Digital signals can be stored in a solid-state memory where they are available for instant retrieval.

The potential advantages of digital transmission are so great that it is widely believed that a digital system will be designated by the FCC for the simulcast channel.

Questions have been raised, however, that must be answered by further field tests[7] before it would be prudent to proceed with a digital transmission standard.

The major question is the extent and reliability of the coverage of stations employing digital transmission. Digital signals do not degrade gracefully, and if the signal-to-noise ratio drops below a threshold as the signal level is reduced, the signal disappears.

The susceptibility of digital transmissions to interference from higher-power analog transmissions in nearby channels may also be a problem. Since it is proposed to use the "taboo" channels for simulcasting, this is a critical criterion.

14.10   Simulcast Channel Format Selection Procedure

The necessity for standardizing the HDTV simulcast channel format requires the FCC to make a choice that will have a large economic impact, since it is assumed that it will be very profitable for the winning company. To make the choice with impartiality and expertise, the FCC appointed an ad hoc committee, the Advanced Television Systems Committee, to study competing proposals, including field testing, and make a recommendation to the FCC for simulcast ATV standards. The tests were scheduled for completion by September 30, 1992, with the decision to come in 1993.

14.11 Proposed Simulcast Channel Formats

The simulcast channel formats that are currently proposed to the ATSC are summarized in Table 14.2. The trend to the digital format is clearly evident from this table; only one of the proposals, Narrow MUSE, employs analog channel encoding. The Narrow MUSE system is described in Sec. 14.11.1. Digital encoding technologies are described in Chap. 4, and the channel encoding systems are described in Sec. 4.11.3.

14.11.1 Narrow MUSE

MUSE is an acronym for multiple sub-Nyquist sampling and encoding. It describes a family of systems in which the signal is subjected to repeated sampling processes, at least one of which is below the Nyquist rate; hence the term sub-Nyquist.

In the parent MUSE system, developed for DBS use by NHK, the bandwidth of the baseband signal is 8.1 MHz, a practical value for satellites. The *RGB* camera component signals are generated in a 1125-line, 60-field, 20-MHz format and sampled at a rate of 48.6 Mbit/s. These samples, in turn, are sampled at one-third this rate, or 16.2 Mbit/s. This reduces the bandwidth requirement by a factor of 3, but a given point on the raster receives a signal only once in every three frames. This is an example of TCI (see Sec. 14.8.4), and the horizontal resolution of moving objects must be restored, in full or in part, by interpolation.

TABLE 14.2   6-MHz Simulcast HDTV/EDTV Formats

System	Proponent	Scanning	Channel coding
Narrow MUSE	NHK	1125/60, interlaced	Analog
DigiCipher	GI, MIT	1050/29.97, interlaced	Digital
DSC-HDTV	Zenith, AT&T	787.5/59.94, sequential	Digital
ADTV	Sarnoff, Thomson, Philips, NBC	1050/29.97, interlaced	Digital
ATVA-P	GI, MIT	787.5/59.94, sequential	Digital

The 16.2-Mbit/s pulse train is converted to an analog signal with an 8.1-MHz bandwidth, which frequency modulates the satellite carrier. The application of MUSE in terrestrial broadcasting systems imposes a maximum bandwidth of 6 MHz. The additional bandwidth reduction is achieved by transcoding the number of scanning lines from 1125 to 750 for transmission. The 1125-line format is restored before display by interpolation.

Narrow MUSE and MUSE 6 are variations of the parent MUSE system that can be transmitted in a 6-MHz channel. MUSE 6 is also compatible with NTSC, but without the requirement for compatibility, Narrow MUSE has slightly better performance.

14.11.2　Digital channel coding

All but one of the systems currently proposed to the FCC use digital channel coding. Their properties are summarized in Table 14.3.

Certain features of these systems should be noted:

1. DSC-HDTV and ATVA-P employ sequential scanning, which requires more bandwidth or a lower number of scanning lines. In the latter case, the number of scanning lines can be restored at the receiver by interpolation, although at some cost and with a slight loss of picture quality.

TABLE 14.3　Operating Parameters, Digital HDTV Systems*

	DigiCipher	DSC-HDTV	ADTV	ATVA-P
Aspect ratio	16:9	16:9	16:9	16:9
Scanning lines	1050	787.5	1050	787.5
Interlace	2:1	1:1	2:1	1:1
Field rate, Hz	59.94	59.94	59.94	59.94
Frame rate, Hz	29.97	59.94	29.97	59.94
Luminance bandwidth, MHz	22	34	24	31
Video sample rate, MHz	51.80	75.34	54	69.05
Video data rate, Mbits/s	13.8	17.2	14.98	15.64
Number of pixels—luminance channel				
Horizontal	1408	1280	1440	1280
Vertical	960	720	960	720
Square	No	Yes	No	Yes
Limiting resolution—luminance channel (TV lines)				
Horizontal	554	504	567	504
Vertical: Static	672	504	672	504
Motion	336	504	336	504

*All systems employ:
• Adaptive quantizers that automatically adjust the quantizing interval in accordance with the amount of detail in the picture that must be transmitted.
• Reed-Solomon forward error correction.
• SMPTE 240M colorimetry.

2. The signal is sampled initially at a rate well above the Nyquist limit. By various bandwidth reduction techniques, the bit rate is brought within the capacity of a 6-MHz frequency band.

3. The vertical resolution is approximately equal to the number of visible scanning lines times the Kell factor. Since the horizontal signal is the result of a sampling process and the location of the pixels is unrelated to the image content, the calculation of the horizontal resolution must include an equivalent of the Kell factor. Specifically [compare with Eq. (2.2)],

$$R_H = \frac{P_H K_f}{A_R} \qquad (14.1)$$

where R_H = horizontal resolution
P_H = number of pixels per scanning line
K_f = Kell factor or equivalent
A_R = aspect ratio

4. The tradeoffs required by sequential scanning are evident from this table. DSC-TV and ATVA-P have sequential scanning lines, but at the expense of more bandwidth and lower horizontal resolution.

5. The resolution of these simulcast systems should be compared with that of the ideal system and a system conforming to SMPTE Standard 240M as shown in Table 14.1. These systems probably approach the performance of these references as closely as possible in a 6-MHz bandwidth.

14.12   Wideband HDTV Systems

The preceding sections have described systems designed to operate in a single 6-MHz channel, the FCC requirement for broadcast simulcasting. Improved performance can be achieved in nonbroadcast applications, where this restriction is removed and wideband systems are permitted, either in a single channel or with an augmentation channel. A variety of modulation formats, both analog and digital, can be used.

The transmission of HDTV programs by wideband satellite circuits has received considerable attention. The MUSE system was originally developed for this purpose. HDMAC (see Sec. 14.13), a variation of the basic MAC format, is another channel coding format for the transmission of HDTV signals.

14.13   HDMAC

HDMAC adds a number of techniques to the basic MAC format (Sec. 3.20) to handle wider baseband signals while achieving bandwidth re-

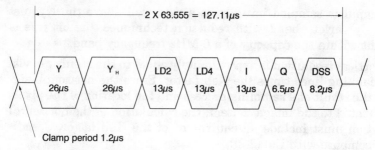

Clamp period 1.2μs

Figure 14.2   HDMAC time sequence.

duction. The time sequence of the component signals during two line scans is shown in Fig. 14.2.

Y and Y_H are the low- and high-frequency components of the luminance signal.

LD2 and LD4 are *line difference* signals; for example,

LD2 = line 2 − (line 1 + line 3)/2

By inspection it can be seen that the LD components are proportional to the line-to-line changes and thus augment the vertical definition.

I and Q are the color difference components.

The video baseband has a bandwidth of 16.8 MHz, which is reduced to 9.5 MHz by the line difference and time compression techniques. The time-compressed signal frequency-modulates the satellite carrier.

14.14   HDTV Equipment

The engineering development of HDTV equipment has proceeded in parallel with the systems development described earlier in the chapter. Although enormous progress has been made, equipment development is just emerging from its infancy, and major technical progress can be expected in the future, particularly after HDTV broadcast standards are established by the FCC. This section provides a brief summary of the status of three major equipment areas, cameras, recorders, and display devices.

14.14.1   Cameras

The imager is the most critical component of an HDTV camera, but a high performance level is required for the lens and the electronic circuitry as well. Difficult tradeoffs must be made in the imager between resolution, signal-to-noise ratio, sensitivity, and lag. Geometric distortion and image registration must be maintained within close tolerances. These same problems exist with NTSC cameras, but their severity is magnified with HDTV. None of them, however, has been sufficiently severe to impede the development of HDTV seriously.

It is probable that the imagers of future HDTV and broadcast cameras will be solid-state devices such as CCDs. It was noted in Sec. 5.3.1 that a typical broadcast camera CCD has 492 rows and 510 columns of pixel photodiodes, for a total of 246,000, and that an HDTV camera CCD would require four times as many, or about 1 million. The higher number has created serious problems, both in manufacturing and in sensitivity. These problems are being solved, however, and an experimental camera has been reported[8] that uses nearly 2 million active photodiodes and exhibits satisfactory sensitivity (2000 lx at f/5.6, comparable with commercial broadcast cameras). The manufacturing cost of cameras of this type will be high (as will be true of all HDTV equipment, at least in the early years), but it appears that the technical problems are being solved.

14.14.2 Recorders

HDTV recorders may include a variety of equipment types:

- Magnetic tape recorders for professional use
- Magnetic tape recorders for use in the home
- Laser and electron beam recorders for transferring electronic recordings to film
- Laser disks for playback

Professional HDTV magnetic recorders have undergone considerable development and are described briefly in this section.

Inexpensive magnetic HDTV recorders for use in the home are largely in the development stage. The problem of designing a low-cost recorder that satisfies the bandwidth and other requirements of HDTV signals is formidable.

Electron beam and laser recording are beyond the scope of this volume. Laser disks for playback only may be more promising than magnetic recorders for home HDTV players.

Either analog or digital recording can be used for HDTV, but multigenerational deterioration of the signal-to-noise ratio and other parameters in analog recordings is severe. Digital recording is often used, therefore, for editing and other applications requiring more than one generation.

Recording parameters and standards based on the SMPTE 240M Standard are frequently used for HDTV recording, and reel-to-reel recorders

TABLE 14.4   HDTV Recording Parameters (Based on SMPTE Standard 240M)

Signal format	Type C band-width	HDTV analog bandwidth	HDTV digital sample rate
Composite	4.2 MHz		14.24 MHz
Luminance Y		30 MHz	74.25 MHz
$R - Y$		15 MHz	37.125 MHz
$B - Y$		15 MHz	37.125 MHz

Parameter	Type C	HDTV
Tape width, in	1	1
Tape speed, in/s	9.6	31.7
Writing speed, in/s	25.59	51.5

have been constructed with the same diameter recording drum as Type C (Fig. 6.9). The parameters for one model are shown in Table 14.4.

For digital recording in this model, the signal is sampled at 74.25 MHz (as compared with 13.5 MHz for the D-1 format; see Chap. 6) and the samples are encoded with 8-bit words, giving a data rate of 594 Mbits/s for the luminance channel and 297 Mbits/s for each of the color difference channels, for a total of 1.118 Gbits/s. If this bit stream were recorded on a single channel with a writing speed of 2027 in/s and a bit rate of 1.118 Gbits/s, the spacing between bits would be only 1.81 μm, which is beyond the capability of current (1992) technology. The solution is to provide 16 tracks, with 1 bit in 16 recorded on each with spacings of 30 μm or 0.030 mm. This spacing requires microscopic gaps on the heads and high-coercivity tape, but it is well within the limits of current technology.

The 16 tracks are recorded by placing groups of four recording heads at 90° intervals around the recording drum so that all can record simultaneously. The result is 16 tracks in groups of 4, each group having been recorded by the heads at a single location. It is apparent that a high degree of mechanical precision is required in the construction of the drum and the placement of the heads.

14.14.3   Display devices

The display device, at the present time usually a kinescope or projection system, is the final equipment component in an HDTV system. Its performance and cost are critical because the complexity and sophistication of program production equipment is wasted without a satisfactory display device that can be sold at a consumer-price level.

Until recently, kinescopes have enjoyed overwhelming preference as display devices for broadcast receivers and for demonstrating HDTV systems. Because of their weight, volume, and cost, however, there is a prac-

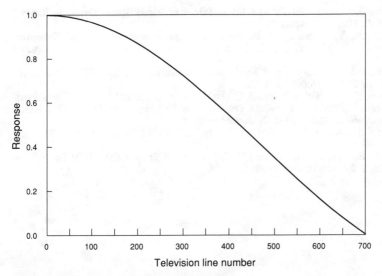

Figure 14.3   Projection TV set aperture response.

tical limit to their size (the limit is not a precise number, but it is of the order of 30 in or less), and, with the growing demand for larger screens, projection systems are becoming more popular.

Current models of projection systems fall short of the requirements for high-quality HDTV systems, with respect to both their brightness and their aperture response. It should be noted, however, that rapid progress continues to be made in their performance.

Figure 14.3 shows the aperture response of a high-quality projection system of current design. It has a limiting resolution of 700 lines and an N_e of 330 lines. This is excellent in comparison with earlier models, but it is below the standards desired for HDTV. It should be compared with Fig. 2.11, which shows the aperture response of a typical direct-view shadow-mask kinescope.

In summary, HDTV systems have the choice of direct-view kinescopes that give excellent performance but are costly and limited in size or projection systems that are marginal with respect to aperture response and brightness. An invention of a better viewing device is needed for HDTV to achieve its full potential, preferably a flat-screen display that will combine high brightness with a high aperture response.

References

1. Mark Shubin, "High Definition Glossary," supplement to *Videography* magazine.
2. L. E. DeMarsh, "HDTV Production Colorimetry," *SMPTE J.*, 100(10), October 1991.

3. Lawrence J. Thorpe, "HDTV and Film—Issues of Video Signal Dynamic Range," *SMPTE J.*, 100(10), October 1991.
4. T. Fukinuki, Y. Hirano, and H. Yoshigi, "Experiments of Proposed Extended-Definition TV with Full NTSC Compatibility," *SMPTE J.*, 93:923–929, October 1984.
5. M. A. Isnardi, "Exploring and Exploiting Subchannels in the NTSC Spectrum," *SMPTE J.*, 97:525–532, July 1988.
6. Ninomiya, Y. Ohutsauka, Y. Izumi, C. Goshi, and Y. Iwadate, "An HDTV Broadcasting System Utilizing a Bandwidth Compression Technique—*MUSE*," *IEEE Trans. on Broadcasting*, BC33:130–160, December 1987.
7. William F. Schreiber, "A Critique of Purely Digital TV," *IEEE Spectrum*, 28, April 1991.
8. Y. Ide, M. Sasuga, N. Harada, and T. Nishizawa, "A Three-CCD HDTV Color Camera," *SMPTE J.*, 99(7), July 1990.

Bibliography

General

ANSI/EIA/TIA, *EIA/TIA Standard, Electrical Performance for Television Transmission Systems*, Electronic Industries Association, Washington, D.C., 1990.

Benson, K. Blair, and J. Whittaker, *Television Engineering Handbook* (rev. ed.), McGraw-Hill, New York, N.Y., 1992.

Fink, Donald G., *Television Engineering Handbook*, McGraw-Hill, New York, N.Y., 1957.

Fink, Donald G., and Donald Christiansen, *Electronic Engineers' Handbook* (3d ed.), McGraw-Hill, New York, N.Y., 1989.

Inglis, Andrew F., *Electronic Communications Handbook*, McGraw-Hill, New York, N.Y., 1988.

Rorabaugh, C. Britton, *Communications Formulas and Algorithms*, McGraw-Hill, New York, N.Y., 1990.

Shannon, Claude E., and W. Weaver, *The Mathematical Theory of Communications*, University of Illinois Press, Urbana, IL., 1949.

Various authors, *Television Technology in the 80's*, SMPTE, White Plains, N.Y., 1981.

Various authors, *Television Technology: A Look Toward the 21st Century*, SMPTE, White Plains, N.Y., 1987.

Various authors, *Television-Merging Multiple Technologies*, SMPTE, White Plains, N.Y., 1990.

Various authors, *Television Technology in Transition*, SMPTE, White Plains, N.Y., 1988.

Various authors, *Tomorrow's Television*, SMPTE, White Plains, N.Y., 1982.

Chapter 1

Fink, Donald G., *Television Engineering*, McGraw-Hill, New York, N.Y., 1952.

Fink, Donald G., and D. M. Luytens, *The Physics of Television*, Anchor Books, Garden City, N.Y., 1960.

Zworykin, V. K., G. A. Morton, and L. E. Flory, *Television* (2nd edition), John Wiley, New York, N.Y., 1954.

Chapter 2

Schade, Otto H., Jr., "Electro-Optical Characteristics of Television Systems," *RCA Review*, Vol. 9, March, June, September, December 1948.

Schade, Otto H., Jr., *Image Quality, A Comparison of Photographic and Television Systems*, RCA Laboratories, Princeton, N.J., 1975.

Various authors, *Better Video Images*, SMPTE, White Plains, N.Y., 1989.

Chapter 3

Hirsch, C. J., W. F. Bailey, and B. D. Loughlin, "Principles of NTSC Compatible Color Television, *Electronics*, Vol. 52, February 1952.

Loughren, A. V., "Recommendation of the National Television System Committee for a Color Television Signal, *J. SMPTE*, Vol. 60, April, May 1953.
McIlwain, K., and C. E. Dean, *Principles of Color Television*, John Wiley, New York, N.Y., 1956.
Various authors, *Television Image Quality*, SMPTE, White Plains, N.Y., 1984.
Wentworth, John, *Color Television Engineering*, McGraw-Hill, New York, N.Y., 1954.

Chapter 4

Luther, Arch C., *Digital Video in the PC Environment*, McGraw-Hill, New York, N.Y., 1991.
Various authors, *Digital Video, Vols. 1–3*, SMPTE, White Plains, N.Y., Vol 1 (1977), Vol 2 (1979).
Watkinson, John, *The Art of Digital Video*, Focal Press, London, 1990.

Chapter 5

Sadashige, K., "An Overview of Solid-State Sensor Technology," *SMPTE Journal*, February 1987.
Thorpe, L., E. Tamura, and T. Iwaski, "New Advances in CCD Imaging," *SMPTE Journal*, May 1988.

Chapter 6

Felix, M., and H. Walsh, "FM Systems of Exceptional Bandwidth," *Proc. IEEE*, Vol. 112, Sept. 1965.
Filbush, D. K., "SMPTE Type C Helical Recording Format," *J. SMPTE*, Vol. 87, 1978.
Iguchi, T., "The SMPTE D-1 Format and Possible Scanner Configurations," *SMPTE Journal*, February 1987.
Robinson, J., *Video Tape Recording Theory and Practice*, Hastings House, New York, N.Y., 1978.
Various authors, *4:2:2 Digital Video*, SMPTE, White Plains, N.Y., 1989.
White, G., *Video Recording and Replay Systems*, Butterworth, London, 1972.

Chapter 7

Sadashige, Koichi, "Developmental Trends for Future Consumer VTR's," *J. SMPTE*, Vol. 93, December 1984.

Chapter 9

FCC Rules, *Sec. 73.600*.

Chapter 10

Eugene R., Bartlett, *Cable Television Technology and Operations*, McGraw-Hill, New York, N.Y., 1990.

Chapter 11

Ha Tri, T., *Digital Satellite Communications* (2nd edition), McGraw-Hill, New York, N.Y., 1990.
Inglis, Andrew F., *Satellite Technology, An Introduction*, Focal Press, Stoneham, MA, 1991.

Miya, K., *Satellite Communications Technology*, KDD Engineering and Consulting, Inc., Tokyo, 1981.

Chapter 12

Kao, C. K., *Optical Fiber Systems: Technology, Design, and Applications*, McGraw-Hill, New York, N.Y., 1982.
Midwinter, J. E., *Optical Fibers for Transmission*, John Wiley & Sons, New York, N.Y., 1979.
Personick, S. D., *Fiber Optics Technology and Applications*, Plenum Press, New York, N.Y., 1985.

Chapter 14

Baldwin, John L. E., "Enhancing Television—an Evolving Scene," *SMPTE Journal*, May 1988.
Benson, K. Blair, and Donald G. Fink, *HDTV-Advanced Television for the 1990s*, McGraw-Hill, New York, N.Y., 1991.
Glenn, William E. and Karen G. Glenn, "High-Definition Transmission, Signal Processing, and Display," *SMPTE Journal*, July 1990.
Ide, Y., M. Sasuga, N. Harada, and T. Nishizawa, "A Three-CCD HDTV Camera," *SMPTE Journal*, July 1990.
Ng, Sheau-Bao, "A Digital Augmentation Approach to HDTV, *SMPTE Journal*, July 1990.
Various authors, *Television Image Quality*, SMPTE, New York, N.Y., 1984.

Index